The Institute of Biology's
Studies in Biology no. 124

Light and Plant Life

Jean M. Whatley

B.Sc., Ph.D.
Department of Botany
Oxford University
Oxford

F. R. Whatley, F.R.S.

First published 1980
by Edward Arnold (Publishers) Limited
41 Bedford Square, London WC1 3DQ

British Library Cataloguing in Publication Data

Whatley, Jean M
Light and plant growth. – (Institute of Biology. Studies in biology ISSN 0537-9024).
1. Plants, Effect of light on
2. Growth (Plants)
I. Title II. Whatley, F R III. Series
581.3'1 QK757

ISBN 0-7131-2785-6

Printed and bound in Great Britain at
The Camelot Press Ltd, Southampton

General Preface to the Series

Because it is no longer possible for one textbook to cover the whole field of biology while remaining sufficiently up to date, the Institute of Biology has sponsored this series so that teachers and students can learn about significant developments. The enthusiastic acceptance of 'Studies in Biology' shows that the books are providing authoritative views of biological topics.

The features of the series include the attention given to methods, the selected list of books for further reading and, wherever possible, suggestions for practical work.

Readers' comments will be welcomed by the Education Officer of the Institute.

1980

Institute of Biology
41 Queen's Gate
London SW7 5HU

Preface

The importance of light in biology cannot be questioned. Its conversion into chemical energy in photosynthesis is fundamental not only to plants but to all living organisms. Light, however, has many other controlling influences on the growth of plants. These can be considered in terms of a number of different properties, for example intensity, spectral distribution and duration. Light intensity is particularly important not only in its conversion to chemical energy but also in some morphogenetic effects and in determining plant distribution. The spectral quality of light is important in phototropism and in the control of germination and flowering. Light duration is important in those photoperiodic effects which control patterns of plant development including those which enable the plant to make use of suitable climatic conditions and avoid unsuitable ones. These effects of light all depend on its absorption by particular pigments, like chlorophyll and phytochrome. The conversion of light into chemical energy involves the use of large amounts of light as a metabolic substrate, whereas its use in regulating plant growth is essentially catalytic. This book attempts to draw these divergent strands together.

Oxford, 1980

J. M. W. and F. R. W.

Contents

1 Light and Plant Pigments: Introduction and Chlorophylls

Visible light represents that part of the electromagnetic radiation spectrum which lies between about 380 nm (which we recognize as violet) and 750 nm (which we recognize as very dark red). This light is the energy source on which plants and hence all other living things depend. Ultraviolet and infra-red radiation, which lie outside of the visible range, may also be of biological importance. In addition to acting directly as an energy source, visible light plays an important regulatory role in the life of a plant. It may be concerned, for example, in the orientation of the plant with respect to the light source, or with the seasonal or daily timing of plant growth. Is there a fundamental process underlying all of these phenomena? How is light detected and utilized?

Radiation must be absorbed by a chemical substance if it is to produce an effect. The subsequent behaviour of the chemical after absorbing light will determine its use by the plant. The absorbing substances (pigment systems) are molecules which contain a chromophoric group responsible for their colours. A compound appears coloured because it absorbs only some wavelengths of white light. The chlorophylls look green because they absorb most of the red and blue part of the spectrum and transmit the green, while β-carotene appears orange because it absorbs well towards the blue end of the spectrum and transmits yellow and red. In biological systems there are only a few types of molecular structure able to absorb light and these are used in precisely controlled ways. The important plant pigments are chlorophylls, phytochrome, flavin, carotenoids and anthocyanins. On the other hand, many organic compounds absorb in the more energetic ultraviolet region, and absorption of ultraviolet light may be associated with various uncontrolled and biologically undesirable reactions, including the breaking of chemical bonds.

1.1 Absorption of light

Three different properties of light may separately affect the metabolism and development of a plant: (*a*) its spectral quality, (*b*) its intensity and (*c*) its duration. The response produced depends initially on the receptive pigment, which determines the wavelengths of light which are absorbed, and secondarily on the intensity and/or duration of illumination. For some purposes (e.g. photosynthesis, which uses chlorophyll) a large scale conversion of light into chemical energy is required and the rate of photosynthesis may increase over a considerable range of light intensity. For other purposes (e.g. photomorphogenesis, which uses phytochrome), only a small scale conversion is necessary to produce appropriate messages or signals, and light of low intensity may be sufficient. In yet other systems, the duration of illumination over a 24 h period (the basis of photoperiodism, which also uses phytochrome) may be important.

What happens when a pigment absorbs light? Either a photoreaction takes place and this leads to energy conservation, or the energy is dissipated in a variety of ways and is no longer directly available to the organism.

We must here refer to Planck's quantum theory of radiation transfer, which states that the transfer of radiation takes place in discrete packets of energy called quanta. The energy transferred (E) is related to the frequency (ν) of the radiation: $E=h\nu$, where h = Planck's constant. When extending this to visible light, Einstein used the term photon for the energy of a single quantum of light. The energy of the photon depends on the wavelength of the light. Frequency is inversely related to the wavelength (λ) so that the energy of the photon is $E=hc/\lambda$, where c is the speed of light ($3\times10^8 ms^{-1}$). Thus a photon of blue light (450 nm) contains more energy than a photon of red light (650 nm). One mole of a compound absorbs N photons, having an energy $E=Nh$, where $N=6.023\times10^{23}$ (Avogadro's number). This total energy absorbed per mole is called an einstein (E). An einstein of red light has less energy than an einstein of blue light, though both have the same number of quanta, N It is often useful in biology to express the amount of light used in a reaction in einsteins, since this shows the actual number of quanta used by the photoreaction.

When a pigment like chlorophyll absorbs light, only wavelengths corresponding to particular electronic states are potentially useful. A quantum of red light ('low energy') supplies the energy needed to move one π-electron from its orbital in the ground state in the ring structure of chlorophyll to the next higher orbital, the first excited singlet state. A

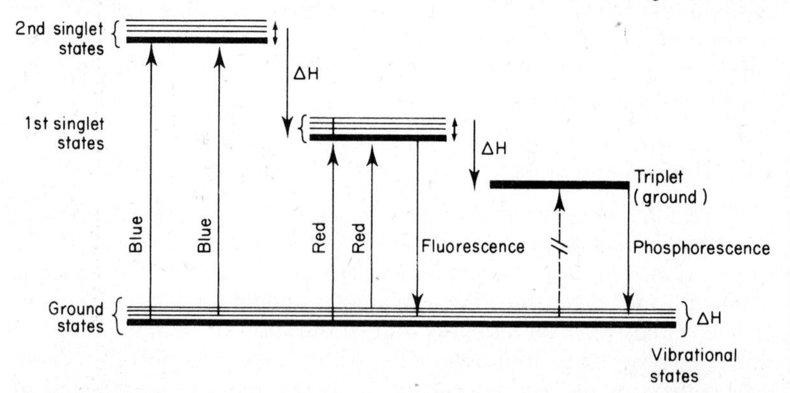

Fig. 1–1 Electronic states in a chlorophyll a molecule. The energy level corresponding to the basic electronic transition is shown as a heavier line, the vibrationally excited sub-states as thin lines.

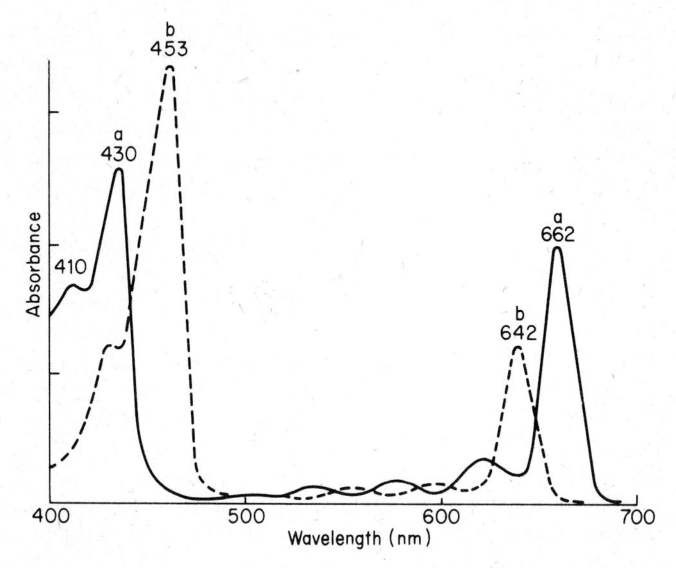

Fig. 1–2 Absorption spectrum of chlorophylls a and b.

quantum of blue light ('high energy') supplies the energy needed to move an electron from the ground state to the second excited state (see Fig. 1–1). A quantum of green light, which has an intermediate amount of energy, does not correspond to the energy needed for any of the various electronic transitions possible in the chlorophyll molecule and chlorophyll cannot therefore absorb green light. This ability to absorb different quanta selectively is represented in Fig. 1–2 which shows the typical absorption spectrum of pure chlorophyll a and of chlorophyll b.

In nature the chlorophyll a is always associated with a protein. This results in some 'fine tuning' of the electron orbitals by interaction between the components of the chlorophyll–protein complex, so that the absorption spectrum of chlorophyll a in the plant is not identical with that shown in Fig. 1–2, though it has the same basic characteristics.

1.2 Fate of excited electrons

An excited electron may participate in a photochemical reaction and this is its most important use in biology. The light energy represented by the excited electron may occasionally provide only the activation energy needed for a reaction to proceed (photosensitization) or part of the energy may be retained in the chemical energy of the products (e.g. photosynthesis, see p. 8). However, only a small percentage of solar radiation can be used for photochemistry – the remainder is dissipated as heat, which may be used locally to drive transpiration, or, on a global scale, may influence climate.

If the excited electron is not used in a photochemical reaction it may return to the least energetic state (the ground state) in a number of ways in which chemistry is not involved. The simplest is by emission of a photon, seen as fluorescence. In chlorophyll this takes place rapidly (10^{-9} s after excitation) and almost exclusively from the 1st singlet state. The wavelength of the fluorescence is slightly longer than that of the least energetic red light needed to activate the electron to the 1st singlet state. Blue light raises an electron to the 2nd singlet state but the excited electron cannot remain in this higher unstable orbit and it very quickly (10^{-10} to 10^{-12} s) falls to the 1st excited state with the emission of some energy as heat. The fluorescence following blue light absorption is therefore at the same wavelength as that following red light absorption. The rapid fall of electrons from the second to the first excited singlet state also means that the photochemically useful energy available from a quantum of blue or red light absorbed by chlorophyll is the same. Energy not lost from the molecule by photochemistry or fluorescence may be lost by internal vibrations or by molecular collisions, and the activated electron may return to the ground state in a number of steps, each accompanied by the loss of heat.

Alternatively the energy of the activated electron may be lost by interaction with another molecule which thereby becomes raised to an activated state. In dilute solution this requires emission of a quantum of fluorescence (photon) by the donor molecule and its reabsorption by the second molecule. This rarely happens. In more concentrated solutions resonance transfer may occur. The acceptor molecule (receiver) may respond efficiently to the electrical field of the energy donating molecule (transmitter) and transfer energy with no intermediate fluorescence. This efficient resonance transfer requires adequate concentration and packing of the molecules such as may occur in the chloroplast membrane and is the likely basis for the co-operation of chlorophylls as light absorbing antenna pigments.

1.3 Action spectra

One technique that has been used to search for the light receptor pigment responsible for a biological process is to determine the action spectrum of that process. Such a spectrum is a graph of the relative effectiveness (action) of light of different wavelengths needed to bring about a biological response, for example, oxygen evolution in photosynthesis, or activation of germination. On the supposition that each biologically effective quantum was absorbed by a receptor pigment (like chlorophyll or phytochrome) and that in the resonating system of the molecule all activated electrons generated by photon absorption are eventually equivalent photochemically, it is reasonable to suppose that the action spectrum will mirror the absorption spectrum of the receptor pigment. To construct an action spectrum it is necessary to determine how many quanta must be supplied at different wavelengths to bring about a standard response. This is done experimentally by measuring dose–response curves. The dose has to be measured in einsteins (because this allows us to ignore the different energies of quanta at different wavelengths) and normally a series of doses has to be supplied before adequate data can be obtained to estimate the quanta required. The most effective wavelengths bring about the standard response with fewest quanta supplied. A plot of relative effectiveness (reciprocal of dose required for response) against wavelength gives the action spectrum. Sometimes it is not practicable to carry out this lengthy series of dose–response measurements. One may then choose a few wavelengths of monochromatic light and supply these at comparable energies (E or μW) to the system under investigation. The energies may be measured in einsteins (E) or microwatts (μW); however, these units are not interconvertible. In this way the response is measured only at a few selected points and the spectrum is therefore incomplete. This is a less satisfactory procedure but may still give strong preliminary indications of the receptor pigment involved.

The action spectrum for the production of chlorophyll in etiolated seedlings is shown in Fig. 1–3 where it is compared with the absorption spectrum of protochlorophyll. The formation of chlorophyll appears to result from a direct photomodification (reduction) of protochlorophyll.

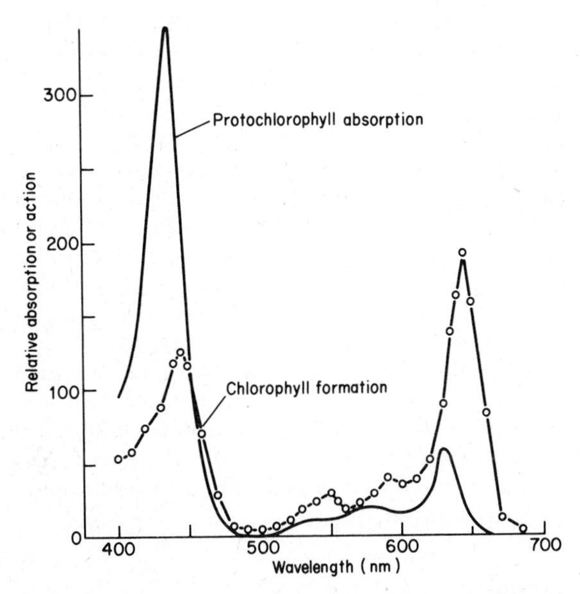

Fig. 1–3 The light stimulation of chlorophyll formation in corn leaves has an action spectrum fairly similar to the absorption spectrum of protochlorophyll. (From KOSKI, V. M., *et al.* (1951). *Arch. Biochem. Biophys.*, 31, 1.)

In general the action spectrum matches the absorption spectrum of protochlorophyll quite well, suggesting a very direct relationship between quantum absorption and photochemical effect. However, light at 440 nm (the blue peak for protochlorophyll absorption) is less effective than expected, probably due to masking effects of the carotenoids which absorb part of the quantum flux so that less is available for absorption by protochlorophyll, prior to its photoconversion to chlorophyll. The enhanced effectiveness of light at 640–660 nm may be due to a greater contribution of photosynthesis at these wavelengths, which could increase the rate of formation of more protochlorophyll and result in an apparent greater effectiveness of red light. Finally, we must note that in the cell the absorption of protochlorophyll, in association with its specific protein, is moved some 10–20 nm further to the red than is indicated by its absorption after extraction into methanol (Fig. 1–3), thus making direct comparison more difficult. The agreement between an action spectrum and the absorption spectrum of the pigment it indicates may therefore not be very close, either for the sort of reasons noted above or because other unspecific absorption of light occurs as it passes through the plant tissue. Action spectra must therefore be treated with some reservations, and may not be taken as more than indicators of the types of pigments which should be further investigated by biochemical and other techniques. It is impossible on the basis of action spectra alone to tell whether phototropism is sensitized by carotenoids or flavins, since both groups of compounds absorb at the blue end of the spectrum and any carotenoids must be expected to seriously shade the flavins.

1.4 The plant pigments

There are three main groups of pigments associated with the principal photoresponses in plants; (1) the chlorophylls, concerned with photosynthesis (2) phytochrome, concerned with some morphogenetic changes, daylength perception and probably also with the phasing of the daily rhythms affecting some plant movements and (3) β-carotene or flavins, concerned with phototropism. Figure 1–4 shows the action spectra of a number of these processes.

Chlorophyll synthesis peaks in both the red and the blue with little activity in the green, as already described. Phototropism peaks in the blue, and red light is totally ineffective. Various photomorphogenetic effects peak in the red at about 660 nm, but the effects are reversed by far-red light at about 730 nm. This reversible type of response suggests the involvement of phytochrome.

1.5 Chlorophylls

Chlorophylls a and b, together with some carotenoids, function as antenna pigments to capture the light energy needed for photosynthesis.

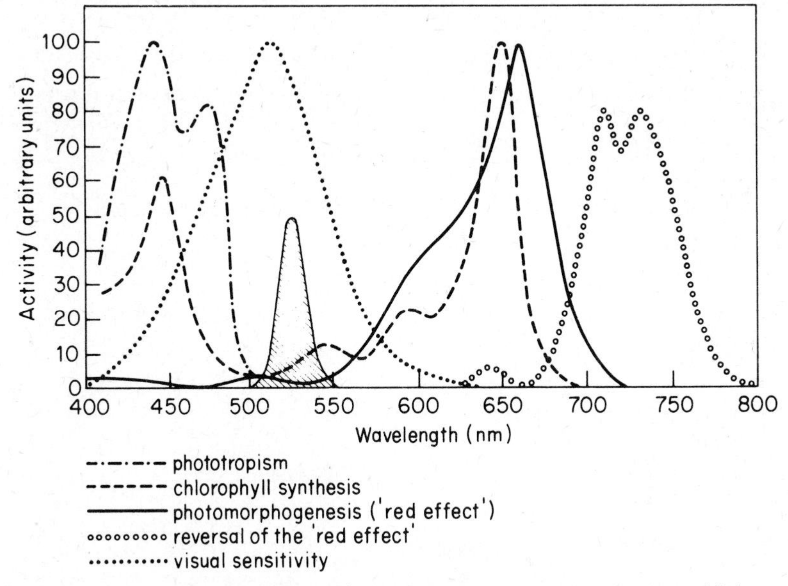

Fig. 1–4 The spectral activity of some photobiological processes; redrawn from data of many authors.

Chlorophyll a is a cyclic tetrapyrrol (Fig. 1–5) with a flat porphyrin head, containing at its centre a covalently bonded magnesium atom. There is a long-chain unsaturated alcohol, phytol, attached as a tail to ring IV and this provides a non-polar anchor which is important in stacking the chlorophyll a molecules in their proper orientation in the chloroplast lamellae. The system of rings in the porphyrin head has nine double bonds in a conjugated system, and these alternating single and double bonds provide many delocalized electrons which can take part in the absorption of light as we have described. The structure of chlorophyll b (the accessory pigment) is very similar, except that the methyl (CH_3–) group on ring II is converted to an aldehyde (–CHO) group, and this slightly alters the position of the two main absorption bands by altering the distribution of the electrons (Fig. 1–3).

Protochlorophyllide, an intermediate in the synthesis of these chlorophylls, is also a cyclic tetrapyrrol but lacks the phytyl side-chain, and has two fewer hydrogens on ring IV than does chlorophyll a. The photoconversion of protochlorophyllide to chlorophyllide a is thus a photoreduction. Only after chlorophyll has been produced can photosynthesis occur.

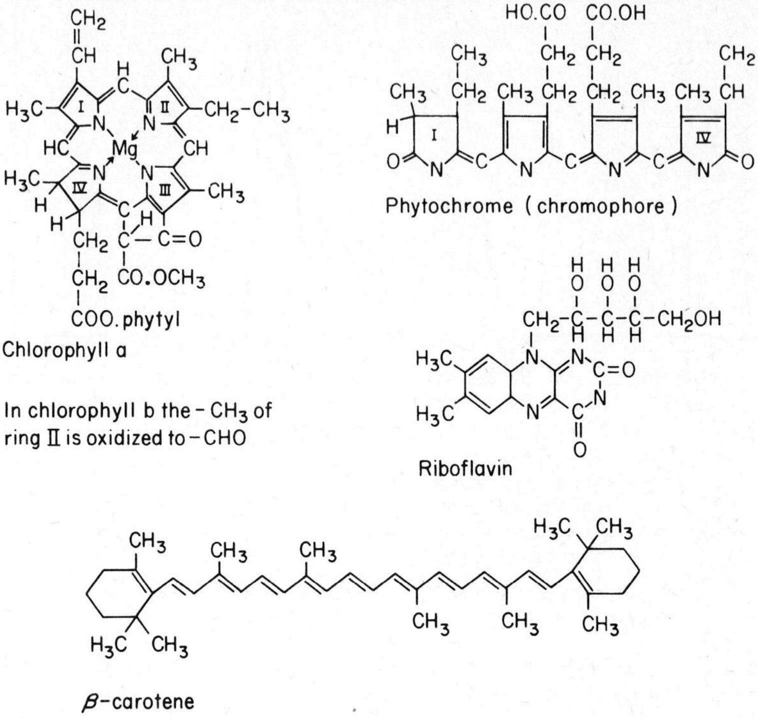

Chlorophyll a

In chlorophyll b the $-CH_3$ of ring II is oxidized to $-CHO$

Phytochrome (chromophore)

Riboflavin

β-carotene

Fig. 1–5 Some photochemically important plant pigments.

1.6 Biochemistry of photosynthesis

Photosynthesis is the process by which plants synthesize organic compounds from inorganic raw materials at the expense of sunlight. Thus light energy is converted to chemical energy, which is eventually stored as carbohydrates and other constituents of plant tissues, and oxygen is evolved as a waste product.

Photosynthesis takes place in chloroplasts, which are present in the green parts of plants. Chloroplasts (Fig. 3–6) are usually discoid organelles, surrounded by a double boundary membrane enclosing a system of flattened membrane sacs, the lamellae (or thylakoids), which are supported in a more or less viscous matrix, the stroma. In electron micrographs the lamellar system may appear to be differentiated into long stroma lamellae and small stacked grana lamellae, but there is no clear differentiation of function between the two types of thylakoids.

The light-capturing pigments, chlorophylls a and b, are found in the lamellae, which are the site of the initial energy transduction of photosynthesis; also in the lamellae, the chemicals $NADPH_2$ and ATP are synthesized, thus generating what has been termed 'assimilatory power'. The components of assimilatory power are used in the stroma to fix CO_2 into sugars by supplying the reducing hydrogens (as $NADPH_2$) and extra energy (as ATP) needed to drive the reductive pentose cycle (Calvin cycle). The 'light' reactions occur in the lamellae and the subsequent 'dark' reactions occur in the stroma.

The overall photochemical reaction characteristic of chloroplasts is termed non-cyclic photophosphorylation and proceeds as follows:

$$NADP + H_2O + ADP + P_i \xrightarrow[\text{chloroplasts ferredoxin}]{\text{light}} NADPH_2 + ATP + \tfrac{1}{2}O_2$$

Under some experimental conditions up to two moles of ATP may be formed per $NADPH_2$. The reduction of NADP proceeds virtually to completion. This is at first sight unexpected, since the reduction of NADP (redox potential, E_0, pH7 = −0.34 V) by water (E_0, pH7 = +0.81 V) is thermodynamically unfavourable as a dark reaction. The amount of light energy that must be supplied to bring about the reduction is not less than 230 kJ mol^{-1} and we must further allow 50 kJ mol^{-1} for the synthesis of ATP. Red light (680 nm) is the least energetic wavelength at which photosynthesis is still effective. Its energy equivalent is 190 kJ/einstein, which is obviously too small. The equation for the photochemical reaction thus requires the co-operation of two or more quanta for each molecule of NADP reduced. Experimental determinations with whole plants show that the lowest measurable quantum requirement is 8–10 quanta per O_2 evolved, i.e. per two $NADPH_2$, which suggests that four quanta are actually needed to reduce each NADP. How can this co-operation occur?

1.6.1 Photosystems I and II

A number of electron carriers participating in photosynthesis have been identified in chloroplasts. It is possible to arrange these carriers in relation to their redox potentials in the form of a Z-scheme (Fig. 1–6), in which two photosystems, I and II, are joined by a thermochemical bridge to allow them to operate in sequence. Transfer of electrons along the connecting bridge releases energy which may be coupled to ATP synthesis. The ultimate electron donor is water and the oxidized product, oxygen, is discarded as waste. The most reduced electron acceptor is the iron–sulphur protein, ferredoxin, which in turn reduces NADP in a dark reaction catalysed by a flavoprotein reductase. Plastoquinone, plastocyanin (a copper protein) and cytochromes occupy positions on the connecting bridge. The Z-scheme shows two separate reaction centres, P_{700} and P_{670}, each recognizable spectroscopically because they have absorption maxima at 700 and 670 nm respectively. P_{700} is connected to an antenna array of light-absorbing pigments in which chlorophyll a predominates (photosystem I = PS I), while P_{670} is independently

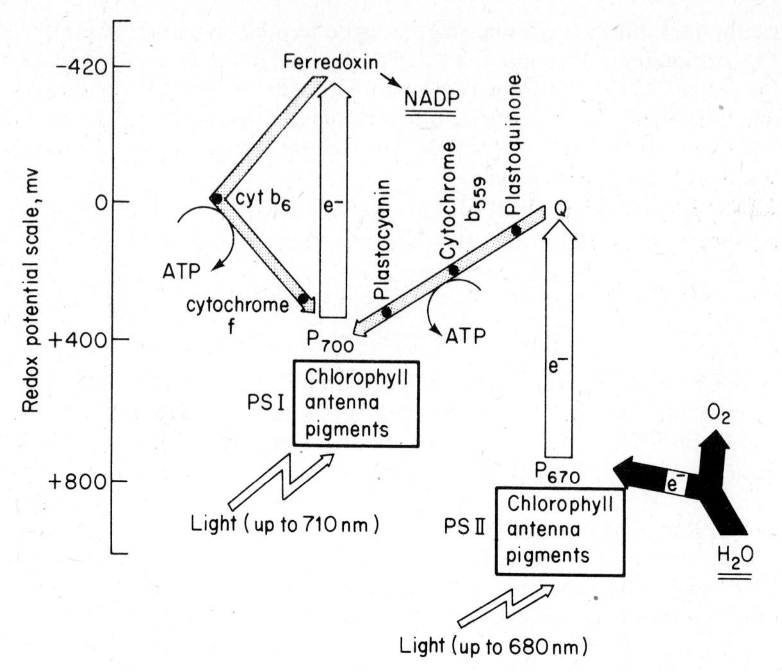

Fig. 1–6 Electron transport sequence in photosynthesis in the form of a Z-scheme.

connected with a light collecting antenna array in which chlorophyll b predominates (photosystem II=PS II). Chlorophyll a molecules in a special arrangement are the eventual receptors of the captured light energy in both P_{700} and P_{670}; thus chlorophyll a is directly involved in both photochemical reactions of photosynthesis.

The first direct evidence for the involvement of two different light reactions in photosynthesis came from the work of R. Emerson in 1943, although it was not so interpreted at the time. If light is absorbed only by the pigments of PS I, as happens above 690 nm in green plants, the efficiency of CO_2 reduction decreases (the red drop). But the efficiency of this absorbed light is enhanced by any additional light taken up by the pigments of PS I (the Emerson enhancement), and an action spectrum for the enhancement shows that chlorophyll b is involved in absorbing the enhancing light in green plants.

A simple biochemical experiment with isolated chloroplasts has been used to separate the two photoreactions. In the presence of the inhibitor, o-phenanthroline, water cannot act as electron donor and both oxygen evolution and the NADP reduction cease. If ascorbate is now added to the inhibited chloroplasts (with a suitable dye to enable ascorbate to reduce a component of the thermochemical bridge) the photoreduction of NADP can now resume at the expense of light energy absorbed by PS I alone; the electrons that would have come from water via PS II are now supplied by ascorbate without the involvement of PS II.

Plastoquinone and ferredoxin can each be readily removed from chloroplasts. In their absence the ability to reduce NADP with electrons from water is lost, but is regained when they are restored, showing they are both essential. The need for other components, which cannot be removed without destroying the chloroplasts, for example cytochrome f, can be shown in other ways. A spectrophotometric assay shows that cytochrome f becomes photo-oxidized in the presence of monochromatic light, absorbed only by PS I. When additional light absorbed only by PS II is provided, it causes a partial reduction of the cytochrome f. Switching off the additional light causes a partial reoxidation of the cytochrome. These observations are readily explained in terms of electrons being 'pulled' from the bridge segment of the electron transport chain by PS I and 'pushed' into the bridge by PS II, the changes in redox state of the cytochrome 'reporting' the balance on the bridge between push and pull.

In a related experiment, two enhancing wavelengths of monochromatic light which gave the same rates of oxygen evolution when applied separately were chosen. On switching from one colour of light to another there is a temporary overshoot or undershoot in the rate of oxygen evolution before the constant rate is resumed, exactly as predicted by the push–pull description above. In addition, if a dark period is interposed between the colour changes, the system appears to 'remember' for many seconds the last wavelength it had received. The 'memory' is a measure of the number of electrons in relatively stable compounds in the thermochemical bridge, which are added to or depleted by the light treatments.

Specific genetic mutants of higher plants and of green algae have confirmed the position of various electron carriers in the Z-scheme. The use of such mutants, in which known compounds have been genetically deleted, are analogous to and confirm biochemical investigations in which known components are removed and replaced, or are inhibited.

Electron paramagnetic resonance measurements of illuminated chloroplasts show clear signals corresponding to the formation of the initial reduced products of the two photosystems. The decay of the signal on turning off the light is biphasic. The 'slow' decay is attributed to plastoquinone (PS II) and the 'fast' to a component associated with ferredoxin (PS I). This experiment again shows that there are two photoreactions, and it has the merit of not interfering with normal electron flow.

A physical separation of the systems responsible for PS I and PS II in chloroplast membrane has also been partly accomplished. Fractions enriched with respect to each of the two photoreactions have been obtained and by adding various combinations of artificial donors and acceptors the partial reactions can be readily shown.

The two photoreactions therefore drive the two consecutive redox reactions, which give rise to the assimilatory power represented by $NADPH_2$ and ATP. However, these initial chemical products of the light transduction can only be generated in very small (catalytic) quantities and are therefore unsuitable for large-scale energy storage. The storage is accomplished in the plant by an extension of the redox sequences into the Calvin cycle. In this cycle each molecule of CO_2 reduced to the level of carbohydrate requires the use of two $NADPH_2$ and three ATP molecules. The final carbohydrate storage product, starch or glucose, in effect stores much of the energy contained in the assimilatory power, and on subsequent oxidation by the mitochondria may release this storage energy to yield ATP and reducing equivalents for further metabolism. The role of these storage products in the throughput of radiant energy into metabolic energy is indicated in Fig. 1–7.

1.7 Physiology of photosynthesis

The amount of photosynthesis carried out by a plant is influenced by the three properties of light – spectral quality, intensity and duration.

1.7.1 Spectral quality of light

The spectral quality of the light is of great importance. An intense source of monochromatic green light at 530 nm would not be absorbed significantly by pure chlorophyll a in solution, so might not be expected to be effective in photosynthesis. This situation may be approached in woodland, where the canopy absorbs blue and red light preferentially and transmits the green, i.e. the understorey plants get greener light (Fig. 1–8). However, the action spectrum for photosynthesis shows that green light is nevertheless effective, though the main peaks are in the red and blue. This appears to be due to absorption of light by the carotenoids, which normally accompany the chlorophylls in the chloroplast membranes; the additional absorbed light is transferred by resonance transfer (see p. 4) to yield excited singlet electrons, equivalent to those produced by direct red light absorption into the chlorophyll antenna arrays. This energy transfer has important implications for the successful growth of understorey plants in forests, which are apparently at less disadvantage than might be expected from the colour of the filtered light.

1.7.2 Intensity of light

The intensity of the light markedly affects the rate of O_2 evolution or CO_2 uptake in photosynthesis. A generalized plot of light intensity (irradiance) against net photosynthesis (measured as O_2 exchange) is shown in Fig. 1–9. The plot shows a linear relationship over a considerable range up to a saturating light intensity, at which point another factor, for example CO_2 concentration, becomes rate-limiting. The length of the linear part of the curve is shorter at lower CO_2

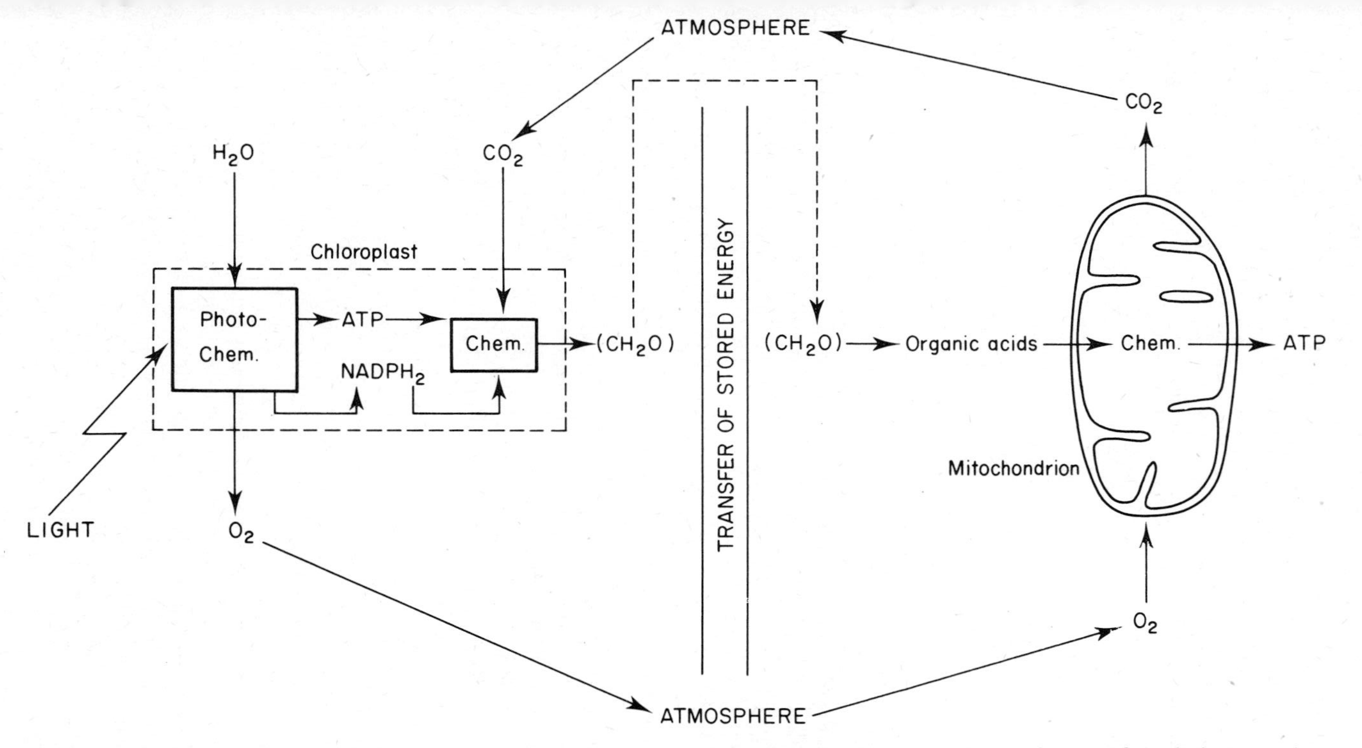

Fig. 1–7 The importance of stored energy compounds (CH_2O = carbohydrates) in the throughput of energy from light into mitochondrial ATP. The transfer of stored energy may be from one organelle to another, or from cell to cell or from organism to organism (food chain). Carbon dioxide and oxygen are recycled in the atmosphere. (From WHATLEY, F. R. (1971). *Proc. Roy. Soc. B.*, **179**, 193–200.)

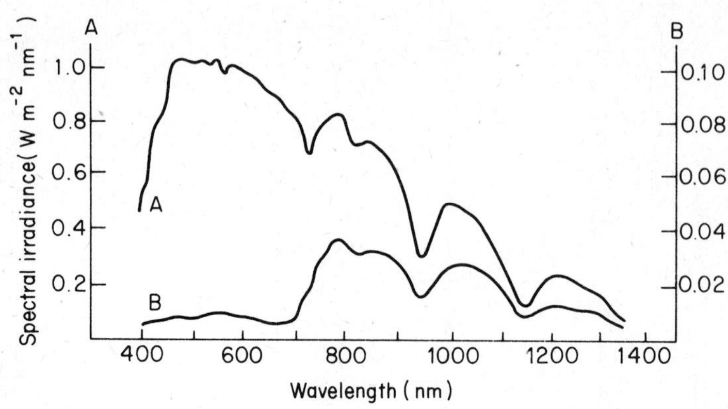

Fig. 1–8 Spectral distribution of sunlight (A) and light beneath a woodland canopy (B). (From FRANKLAND, B. (1976). In *Light and Plant Development* (ed. H. SMITH). Butterworth.)

concentrations. The slope of the initial linear part of the curve indicates the number of quanta needed to evolve each molecule of O_2 (or reduce each molecule of CO_2). The slope is characteristic of the species and depends on physiological factors as well as on the strict quantum requirements of photosystems I and II. On the same plant some leaves (sun leaves) normally receive more light than others (shade leaves). It is characteristic that the sun leaves have a greater quantum requirement than the shade leaves, so that the initial slope for shade leaves is greater. However, the shade leaves become saturated at a much lower light intensity (irradiance) than do the sun leaves. Excess light may damage the photosynthetic apparatus; the leaves may become bleached and the rate of photosynthesis may decline. The light intensity at which this happens depends on the particular species and on whether the leaf is a sun or shade leaf.

At the lowest light intensities the real rate of photosynthesis becomes less than the rate of respiration (which, in a formal sense, is the reverse of photosynthesis) and the net rate of photosynthesis becomes negative. The intensity at which the two processes just balance is called the light compensation point, and this is also characteristic of a particular species. This point may change at different stages of the life cycle. A plant cannot of course survive at less than its compensation point for any length of time, though it may tolerate short periods of intense shading, in addition to being in the dark during part of the 24 h day–night cycle. Plants adapted to grow in shade conditions characteristically have low light compensation points.

A plot of light intensity (irradiance) versus net rate of photosynthesis for any plant will be related to the generalized plot (Fig. 1–9). The graph will give some information about the limits to the plant's distribution and possible habitat imposed by the amount of light it receives.

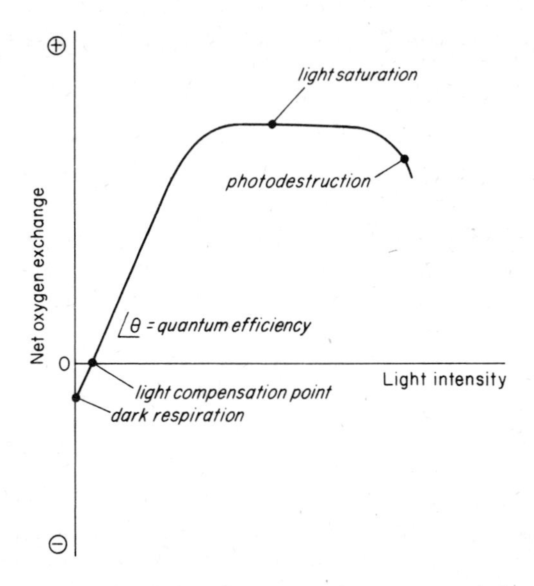

Fig. 1–9 A generalized plot of oxygen exchange against light intensity.

1.7.3 Duration of light

The greater the duration of illumination the more photosynthesis will, in general, be accomplished and if a plant is exposed to longer days it can be expected to fix more CO_2. However, if plants are provided with a prolonged period of light they may not continue to photosynthesize because of a temporary inability of the chloroplasts to store all the extra starch. Also higher temperatures (such as are common at midday in summer) may cause wilting and stomatal closure, thus limiting CO_2 uptake. If light is supplied continuously some plants become bleached and chlorotic. There is, therefore, no absolute relationship between daylength and photosynthetic yield.

1.7.4 Direction of light

There is no evidence that the direction of illumination is important for photosynthesis, although the angle of incidence of leaves with respect to the direction of the light source may be changed by the plant. Where light intensity is high the leaves tend to take up a vertical position so that the minimum light-absorbing surface is exposed to the sun. By contrast, in shady conditions leaves may be positioned horizontally so that the maximum light absorbing surface faces the sun.

A more detailed consideration of the various aspects of photosynthesis is contained in the companion volume of this series by HALL and RAO (1977).

2 Light and Plant Pigments: Other Pigments

Unlike the chlorophylls, the other pigments – phytochrome, β-carotene and flavins – are not involved in the large scale conversion of light into chemical energy. Rather they are concerned with accepting light signals and then creating chemical messages which may change the course of plant metabolism and growth.

2.1 Phytochrome

Phytochrome appears to be associated with many different types of response by the plant. Its mode of action is unclear, but it may act by altering the permeability of membranes and by the activation or repression of key genes which influence protein synthesis. A working model suggesting the photochemical basis of its mode of action is shown in Fig. 2–1.

Phytochrome is a large protein with a molecular weight of about 120 000. The chromophore group responsible for its visible light absorption is a linear tetrapyrrol; there are at least two of these chromophore molecules attached to each phytochrome protein, but the chromophores account for only 1% of the total weight of the phytochrome. The tetrapyrrol is a close relative of the phycocyanin chromophores which, like chlorophyll b in green plants, serve as supplementary light absorbers (accessory pigments) in the photosynthetic blue-green and red algae. Its existence in higher plants may represent a new physiological use for the linear tetrapyrrols developed during the evolutionary sequence from primitive algae to green plants.

The protein of phytochrome contains a high proportion of acidic and basic amino acids and of the sulphur-containing amino acid, cysteine. This indicates that phytochrome is a highly charged and highly reactive

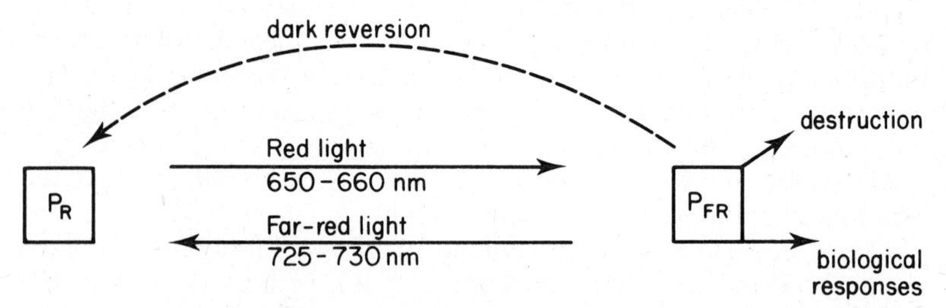

Fig. 2–1 A suggested mode of action of phytochrome. Only phytochrome in the far-red form is thought to give the biological response.

molecule, potentially capable of many internal rearrangements leading to changes in shape. The isolated phytochrome molecule appears to have a dumb-bell shape and treatment with proteolytic enzymes shows that it can be cut across the middle into two equal sub-units, each containing a chromophore.

2.1.1 *P_R and P_{FR}*

On illumination of isolated phytochrome with red light (650 nm) there follows a characteristic spectral change from a blue form (P_R) to a blue-green form (P_{FR}); the reverse change from P_{FR} to P_R occurs on illumination with far-red light (730 nm). The spectra of the two forms are shown in Fig. 2–2. The strong absorption at and below 300 nm results from aromatic rings in some of the amino acid constituents of the protein and not from the chromophoric group. On the basis of these spectral changes it has been suggested that the length of the conjugated system of the tetrapyrrol in P_R (responsible for its absorption spectrum) may be increased on illumination with red light. This is brought about by migration of hydrogen atoms (protons) into suitable accepting groups in an amino acid of the protein part of the phytochrome (Fig. 2–3). As a consequence, the extended conjugated system now present in P_{FR} is 'tuned' to a slightly lower set of electronic energy levels (the shared π electrons are spread over a larger number of resonating bonds) and the absorption bands are modified and moved towards the longer wavelengths.

Measurements have been made of the circular dichroism of phytochrome (a very sensitive method to detect changes in the shape of molecules) as it undergoes the photochemical change from P_R to P_{FR} via a series of 'bleached' intermediates. The evidence suggests that red light

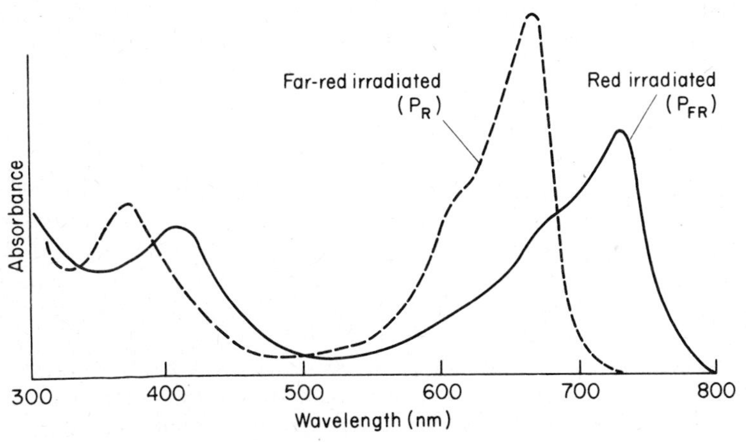

Fig. 2–2 Absorption spectra of the two forms of phytochrome. (From HARTMANN, K. M. (1966). *Photochem. Photobiol.*, 5, 349.)

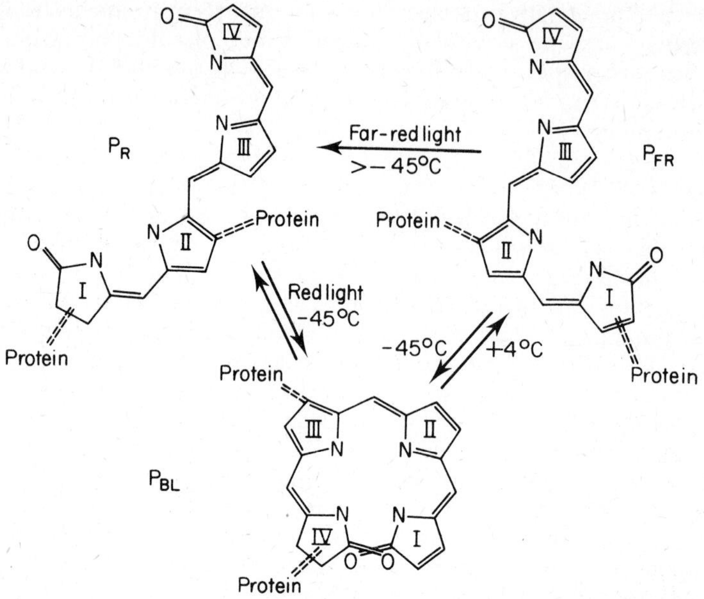

Fig. 2–3 A hypothetical scheme for conformational changes during photo-conversion of the forms of phytochrome. (From BURKE, M. J., *et al.* (1972). *Biochemistry*, **11**, 4025. Reprinted with permission. Copyright by the American Chemical Society.)

causes a very large shape change in the tetrapyrrol from the *cis*-configuration which it has in P_R to the *trans*-configuration in P_{FR}. Figure 2–3 suggests the sort of change of shape involved, and indicates the nature of a possible folded intermediate (P_{BL}). The change indicated would require the loss of a hydrogen atom from the 'bottom' pyrrol ring and its absorption by an ionized group of an amino acid in the phytochrome protein. The absorbed quantum of red light provides the energy needed for this transfer; the reverse transfer receives its activation energy by absorption of a far-red quantum.

If, as shown in Fig. 2–3, there are two points of attachment of the tetrapyrrol structure to the protein (we do not know the details of the groups involved) then the change in relative position of the 'bottom' pyrrol ring during *cis–trans* isomerization means that the actual shape of the protein molecule will be forced to change on illumination. This suggested sequence of events is similar to that which occurs in vision when the chromophore of the better-known visual pigment, rhodopsin, changes from *cis*-retinal to *trans*-retinal, a transformation which is accompanied by conformational changes in the protein structure. The visual protein is associated with the disc membrane in the retina and its conformational change brings about a charge separation on the

membrane which is responsible for the neural signal translated into vision.

The amount of phytochrome present in the plant in the P_{FR} form is usually expressed as P_{FR}/P_{TOTAL}, both of which can, under favourable conditions be measured spectroscopically. It has been suggested that a specific ratio of P_{FR} to P_{TOTAL} must be reached to produce a morphogenetic or biochemical effect. This ratio can be set experimentally by irradiation with a combination of wavelengths of monochromatic light which cause the photoreaction $P_{FR} \rightarrow P_R$ or its reverse. In nature the ratio of red : far-red wavelengths in sunlight is altered as it is filtered through green leaves; thus leaves lower down in a forest or crop canopy receive different ratios of red : far-red light from those at higher levels, and as a consequence the ratio $P_{FR} : P_{TOTAL}$ changes (Fig. 1–8). This ratio change could form the basis of detection of shading and might provide the signal which is later translated into a biochemical or morphogenetic reaction. Implicit in this suggestion is the idea that a particular response is switched on or off when a particular ratio is reached. How is it possible for such a ratio change to give rise to an effective message that will cause any one of the many different photomorphogenetic, biochemical or kinetic effects described in later sections? Can one relate it to conformational changes accompanying the transformation of $P_R \rightarrow P_{FR}$? Does the location of phytochrome in the plant offer any clues?

2.1.2 *Location of phytochrome in the plant*

It is relatively easy to detect phytochrome in non-green (and etiolated) parts of the plant but difficult to show it in the green parts of plants. It has also proved difficult to get good evidence for the intracellular localization of phytochrome. However, if the cell is homogenized and its various membrane components are separated by centrifugation there is a strong indication that phytochrome is associated with various cell organelles and with the plasmalemma. The amount of phytochrome which remains associated with the membranes after centrifugation is greater after pre-treatment of the plant material with red light than with far-red light; this suggests that the association of phytochrome with the various membrane systems may be only temporary and occur only on photoconversion to P_{FR}. Most of the phytochrome, especially P_R, is present in the cytoplasm and not in association with membranes. The large amount of phytochrome in the cytoplasm may correspond to an inactive fraction, while that associated with the membranes may correspond to an active pool.

Some attempts have been made, using a microspectrophotometer, to observe changes in the absorption spectra associated with specific subcellular localities within oat coleoptile cells. Following red or far-red illumination, absorbance changes associated with the nucleus took place at wavelengths approximating the maximal absorption by phytochrome. It was tentatively concluded from these experiments that the nuclear membrane is an important site for phytochrome. There is also evidence for its localization in the plasmalemma. The use of immunological techniques (in which antibody produced in rabbit serum in response to purified phytochrome is allowed to react with thin slices of plant tissues and the antibody-phytochrome produced is 'stained' by a histochemical method) has also indicated that while phytochrome is concentrated in various cell membranes, it is certainly present in the cytoplasm. Finally, there is evidence from the differential responses to polarized or non-polarized red light observed in the filamentous green alga *Mougeotia* (work on chloroplast orientation) that phytochrome molecules are localized near or in the plasmalemma in precisely oriented molecular patterns.

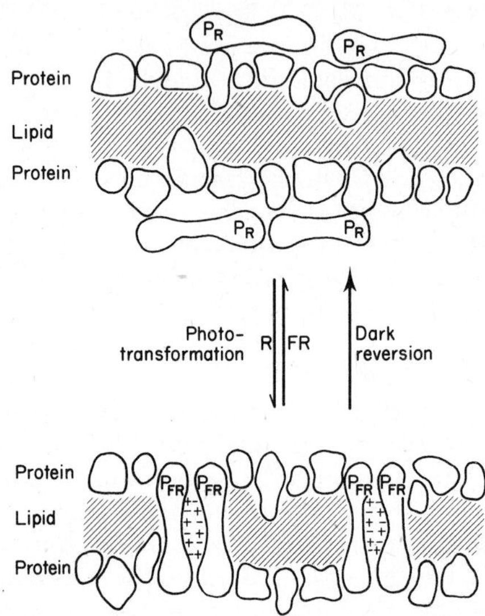

Fig. 2–4 A possible mechanism by which phytochrome could regulate membrane transport with low specificity and thus cause a multiplicity of apparently unrelated effects. (From SMITH, H. (1976). In *Light and Plant Development* (ed. H. SMITH). Butterworth.)

2.1.3 Mechanism of phytochrome response

The problem then is to suggest a common mechanism underlying the very varied phytochrome-mediated responses by the plant to red and far-red light. A possible mechanism, shown in Fig. 2–4, is based on the alteration of shape which the phytochrome molecule undergoes when changing from one form to another. The figure suggests a loose association of P_R with the surface of a selectively permeable membrane, which is impermeable to various water soluble metabolites and hormones. On photoconversion to P_{FR} the phytochrome molecule assumes a new shape with different surface properties. This new molecule now 'dissolves' in the membrane and groups of P_{FR} molecules become aligned across it (self-assembly) to generate aqueous 'pores' permeable to water soluble substances. The permeability characteristics of the membrane would thus be greatly altered as P_{FR} was formed and a number of water soluble components, normally compartmentalized within a cellular organelle inside a particular cell, could become redistributed. The effect produced by this redistribution would then depend on the nature of the water soluble substance released and not directly on phytochrome. This could account for the large number of known phytochrome-controlled responses. The substance released (the message or stimulus) could be a hormone, an ion or a metabolite. The effect of a particular P_{FR}/P_{TOTAL} ratio on the permeability of a membrane would be expected to depend on the precise chemical nature of the membrane itself and a threshold value of P_{FR} needed for the generation of aqueous pores might be expected.

The release of P_{FR} from the membrane requires its reconversion to P_R. This conversion may occur quickly in response to far-red light, as may happen in nature, in say, the detection of shading (see p. 19), or artificially in a night-break experiment, in which far-red light is given during the normal period of darkness. Alternatively the reversion of P_{FR} to P_R may occur slowly in the dark and this spontaneous reversal could act as part of the mechanism for detection of daylength.

Light quality (its spectral distribution) is of considerable significance in the action of phytochrome. It is clear from the action spectra for various photomorphogenetic effects (Fig. 1–4) that only certain wavelengths of light are important in producing a response. Where an effect is caused by red and reversed by far-red light a phytochrome-mediated response is likely.

For growth of plants under artificial conditions the spectral quality of the light sources may be important in addition to the quantum flux (light intensity). Tungsten lamp sources are relatively deficient in blue light and rich in the infra-red, causing problems both in photomorphogenesis and in temperature control in illuminated growth cabinets. The use of fluorescent tubes, while solving some of the problems of temperature control, initially gave additional morphogenetic problems because they were rich in blue and poor in red light. Chemical modification of the tubes has allowed some of the deficiencies of the original tubes to be partially overcome (from the point of view of the plant) and a combination of blue and warm white fluorescent tubes, sometimes supplemented by a few tungsten bulbs, has been successful in producing an acceptable light climate for morphogenesis and photosynthesis for most plants used experimentally.

The *light intensity* needed to cause a phytochrome-mediated response is usually low, though a high intensity response (p. 30) may occasionally be of importance. However, moonlight and starlight apparently fall below the critical intensity needed for any response.

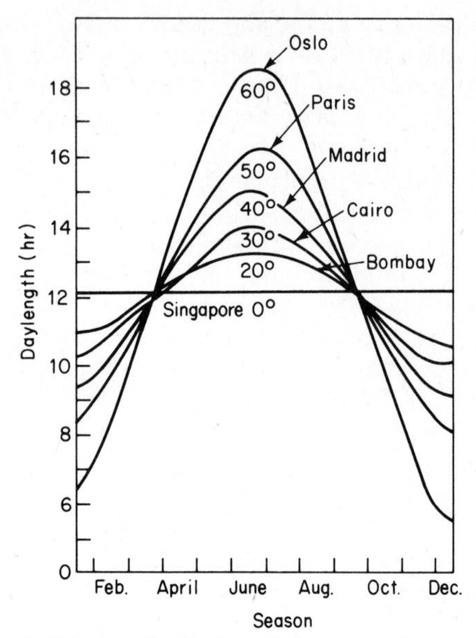

Fig. 2–5 The annual changes in day length, from sunrise to sunset. The amplitude of the changes increases with increasing latitude.

The *duration of illumination within a 24 h cycle* is also of importance in many instances. The plant may use day length to obtain predictive information about the season. Of all the environmental factors regulating plant growth the only truly constant one is the precession of day length (Fig. 2–5). The daylength can be detected by many plants by making use of the phytochrome system. Generally, detection of daylength appears to be of greater importance to plants growing in temperate and high latitudes than in the tropics (p. 61).

2.2 Daily rhythms

Phytochrome is also involved in a number of plant responses associated with the daily rhythm of light and darkness. One of the earliest observations was on the rate of growth of oat coleoptiles, which commonly grow faster in the dark than in the light. However, if the coleoptiles are placed in continuous darkness the growth rate continues to oscillate for some time, and the oscillation is initiated at the time of transfer into the dark. This suggests a diurnal rhythm entrained and reinforced by the light–dark trigger.

2.2.1 *Plant movements*

Another early observation was of the daily movements of leaves to a horizontal position in the morning and to a vertical 'sleep' position in the

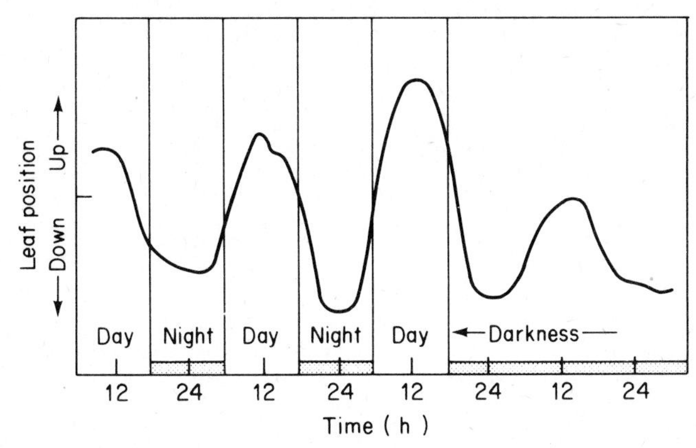

Fig. 2–6 The diurnal movement of *Albizzia* leaves. The leaves are elevated in the morning and lowered in the evening. If kept in the dark movement is repeated for several cycles with decreasing amplitude. (From BÜNNING, E. (1959). *Handbuch Pflanzen-physiol.*, **17**, 8; as drawn by LEOPOLD, A. C. (1964) in *Plant Growth and Development*. McGraw-Hill.)

evening. This regular cycle appears in many species, for example *Albizzia* and *Phaseolus*, to be a phytochrome-controlled phenomenon, with red/far-red reversibility. However, if the plants are placed in the dark after the regular daily cycle is established, they, like the oat coleoptiles, continue to respond for a time as if they were being illuminated and darkened in the normal sequence. There is obviously a diurnal rhythm, again reinforced by the action of light on phytochrome (Fig. 2–6).

The anatomy of plants which have sleep movements is modified to include a pulvinus or hinge pad at the base of the petioles in which the metabolism involved in sleep movements is actually carried out. The pulvinus contains a group of specialized cells which undergo rhythmic changes in water content, apparently in response to redistribution of potassium ions, causing shrinking and swelling responses that result in the sleep movements.

The daily opening of some flowers is also light triggered. Normally light causes opening of the flowers, but in a few plants the flowers, for example *Oenothera* (evening primrose), close during the day and open at night.

Some plants characteristically turn towards the sun (heliotropism). In *Helianthus annuus* (sunflower) the light stimulus is perceived by the leaves closest to the shoot tip. The leaves more distant from the sun get more intense illumination (as shown in Fig. 2–7) and elongation growth on that side of the stem is stimulated by the increased auxin diffusing from the more brightly lit leaf. When the opposite leaf comes into a position where it receives the same amount of light the turning stops. However, the equilibrium is unstable because the sun moves from east to west; but since

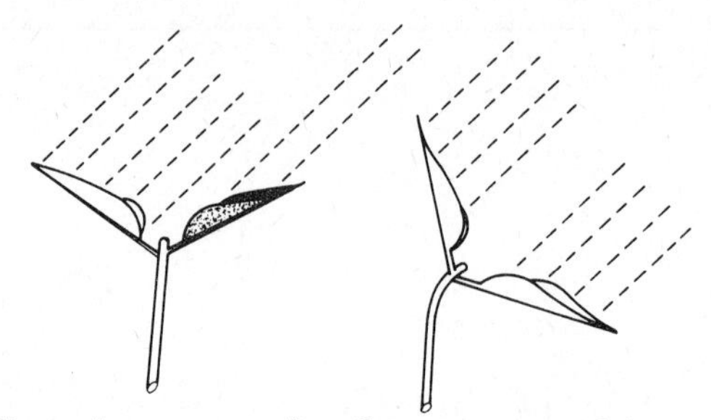

Fig. 2–7 Heliotropic curvature of sunflower. Because of the angle at which the leaves are held the leaf 'further' from the sun receives more light than the other. The bending response equalizes the light on the two leaves. (From SHIBAOKA, H. and YAMAKI, T. (1959). *Sci. Papers of Coll. Gen. Educ., Tokyo*, **9**, 105; as drawn by LEOPOLD, A. C. (1964) in *Plant Growth and Development*. McGraw-Hill.)

the equilibrium is continually restored the plant twists to follow the sun. This diurnal heliotropic rhythm is maintained for several days in complete darkness, after which a light stimulus is needed to re-entrain the system.

2.2.2 *Stomatal movements*

Another daily rhythm is the opening and closing of stomata. In order to get CO_2 into the leaf the stomatal pores must be opened, but this also causes loss of water vapour by transpiration. At night when CO_2 fixation is not possible it is advantageous to limit the water loss, and the stomata close. The stomata begin to open just before dawn and are normally fully open within half an hour of sunrise. The light trigger again appears to entrain this cycle. When the leaf is placed temporarily in the dark the diurnal cycle of opening and closing continues for several days. The stomata open because the guard cells actively take up potassium from the surrounding epidermal cells and consequently swell because of the osmotic uptake of water. The energy for this potassium uptake comes from respiratory metabolism, reinforced by photosynthesis in the chloroplasts of the guard cells. A loss of turgidity follows the loss of potassium ions from the guard cells. If the night is interrupted by a short period of light, the stomata begin to open, but close again when the light is turned off.

In many succulent desert plants, including *Kalanchoe* and the cacti, the cycle is reversed and the stomata open at night for the plant to absorb CO_2 (which is incorporated into the organic acids) and close during the day and so avoid loss of water by transpiration. Light entrains this cycle of events. The organic acids are converted photosynthetically to sugars during the following light period (Crassulacean Acid Metabolism). The phasing of this stomatal sequence is phytochrome controlled.

There are many more diurnal rhythms in plants, some of which are kept accurately cycling on a 24 h period by the entraining effect of a daily light–dark cycle. These include the daily cycle of ion uptake by the roots (related also to transpiration), cell division, respiration and guttation (secretion of water through glands at the leaf tip). Finally there is the obvious phasing of photosynthesis to daylight hours and the subsequent translocation of carbohydrates to roots and storage organs in the dark.

2.3 β-carotene and flavins

Not all plant movements are primarily controlled by phytochrome. Phototropism, the bending of a plant shoot in response to illumination from one side, apparently takes place principally in response to the effect of light on either β-carotene or a flavin, though phytochrome may play some part. Most of the work on phototropism has been done on etiolated (non-green) oat and corn coleoptiles – unfortunately we have almost no experimental evidence about the mechanisms that operate in green shoots.

The *action spectrum* of the bending response in coleoptiles indicates a receptor pigment that absorbs in the blue but not in the red part of the spectrum (Fig 1–4). Two compounds, found in the coleoptile tip, whose absorption spectra resemble this action spectrum are β-carotene and riboflavin. So far it has not been possible to decide with certainty which, if either, of these pigments is the actual receptor. Each pigment has extensive conjugated double bond systems whose π electrons are 'tuned' to accept blue light quanta. β-carotene appears to be relatively inert chemically, although carotenoids have been proposed as effective anti-oxidants which prevent the photodestruction of chlorophyll at normal light intensities. β-carotene is commonly found in association with membranes, but we cannot suggest any plausible redox reaction or *cis–trans* isomerizations to explain how it might work in the perception reaction of phototropism. On the other hand, the illumination of riboflavin in the presence of auxin *in vitro* brings about a rapid photodestruction of the auxin, and it has been argued that this could form a common basis for the action spectrum and for the redistribution of auxin. The chief objections to this otherwise very attractive hypothesis are that the total amount of auxin remaining in the coleoptile after redistribution is the same as the amount present in the dark (so the photodestruction mechanism is not favoured), and that the intensity of light needed for rapid photodestruction is much higher than that needed to produce a phototropic curvature. Although the action of riboflavin in catalysing the photodestruction of auxin was earlier thought to suggest an acceptable mode of action the current view supports the involvement of β-carotene.

As indicated by the action spectrum (Fig. 1–4), phototropic curvature is induced by blue light and not by red. However, pre-illumination of the coleoptiles with red light (660 nm) enhances the phototropic response to blue light and far-red illumination (760 nm) abolished the enhancement.

It looks as if phytochrome also plays a part in regulating phototropic curvature.

The effect of *light intensity* (irradiance) on the bending response of coleoptiles is complex. There are three separate responses to light. It was early found that lateral illumination with white light caused the coleoptile to bend towards the light at the lowest light intensities (maximum response at 0.1 Jm^{-2}), but above this intensity the coleoptile may bend away from the light (maximum response at 10 Jm^{-2}). At still higher irradiances a strong positive curvature towards the light is seen (maximum response at 100 Jm^{-2}). Normal daylight falls in the range that elicits the third response.

The *duration* of illumination required to cause a subsequent bending response in the dark is quite short. About five minutes is enough for a full response at the higher intensities needed for the second positive response (the 'natural' response in daylight), and a shorter period of 2–3 minutes suffices for the first positive response.

The *direction* of illumination appears at first sight to be important in phototropism. However, the coleoptile actually detects the difference of light intensity on the two sides of the tip and 'decides' the direction from this information. If one side of an *Avena* (oat) coleoptile is illuminated by a very fine beam of light from directly above, the resultant curvature (in the dark) is towards the more brightly illuminated side, and is therefore at right angles to the actual direction of illumination.

The unequal lighting is *perceived* most strongly at the tip of the coleoptile. Following decapitation a new area at the end of the truncated coleoptile becomes sensitive to light and a new 'physiological tip' is formed.

Following perception the *induction* of the phototropic response requires the active redistribution of the growth hormone, auxin, which is produced in the tip. The auxin becomes depleted on the side nearer the light source and concentrated on the opposite side. The *bending response* now involves differential growth in the extension zone below the tip as a result of the unequal auxin distribution, and the tip is thus brought into a new equilibrium position with respect to the light source. There is a complex interaction between the phototropic response to light and the geotropic response to gravity; examination of the phototropic response uncomplicated by geotropism will require future experiments in space outside the earth's gravitational field, perhaps in a future Skylab. The importance of the phototropic response is that it carries the newly expanding photosynthetic tissue rapidly away from a shaded position to one with better lighting.

3 Germination and Seedling Establishment

Many aspects of plant development depend on responses to the presence or the absence of light. Some responses depend on the daily rhythm of light and dark, for example photosynthesis and translocation; others depend on seasonal rhythms of day length, for example flowering. Growth of the plant as a whole reflects the complex interactions of these responses in different parts of the plant at different developmental stages.

Light is important throughout the whole of the life cycle of the plant. The successful establishment of an angiosperm seedling may require light (or lack of it) to break dormancy, to promote extension above the soil surface and to produce leaves of a size, shape, orientation and chlorophyll content adequate for efficient photosynthesis. Subsequent changes in plant form (e.g. production of flowers, tubers or bulbs) may require a specific photoperiod. At the end of a growing season, leaf abscission, development of frost resistance or various types of dormancy may be similarly controlled. Any such processes, which assist survival in, or through, periods of unfavourable growing conditions, may be just as important as those which control growth itself.

Germination involves the activation of the previously quiescent or dormant embryo. Seedling establishment involves subsequent development of the germinating seedling to a state where it no longer depends on stored nutrients, but can exist independently by using its own photosynthetic products.

Although a single plant may produce large numbers of viable seeds in one year (e.g. *Epilobium* sp., *c.* 80 000), few seedlings can succeed in getting established. Becoming independent as quickly as possible is a major advantage in the competition to survive. It involves not only the light-dependent development of the photosynthetic apparatus but also a variety of other responses to light associated with metabolism and growth.

3.1 Dormancy break

The seeds of most species can begin to germinate as soon as they are provided with water, oxygen and a suitable temperature. However, for a variety of reasons (see VILLIERS (1975) section 5.7) seeds of other species will not germinate immediately. Seeds in such a state are called dormant and an additional stimulus is necessary before germination can take place. Light (or lack of light) provides the stimulus necessary for breaking one form of seed dormancy; it is only effective after water has been imbibed and it acts by removing a blockage in the metabolic pathway of the embryo.

Under some conditions of stress, seeds of species which normally do not require light to germinate become light-dependent. Under other conditions a normal requirement for light may be lost.

As a result of selection over a long period, seeds of cultivated plants are generally unaffected by light. However, seeds of weed species are very variable in their requirements for light, even within a single batch of seeds produced by one plant. This capacity for variation may have useful ecological consequences by ensuring that at least some seeds may germinate, no matter what light conditions prevail.

The manner in which the light must be presented in order to break dormancy is not always the same. It is convenient to consider the light requirements for germination under three headings: (*i*) trigger effects, requiring only a short period of light of low intensity (low energy system), (*ii*) light inhibition, and (*iii*) the effects of continuous or periodic irradiation (high energy system – a photoperiodic response).

3.1.1 Low energy phytochrome system with light acting as a trigger

Some seeds require very small amounts of light for germination, for example *Lactuca sativa* var. Grand Rapids (lettuce) and *Lythrum salicaria*. Both the percentage and rate of germination depend directly on the quantity of light applied (intensity × duration) until a saturating intensity is reached. Thus in *Rumex obtusifolius* (bitter dock), 100 sec illumination at 1000 lux was found to just saturate the system, i.e. to produce maximum percentage germination, while a similar effect was produced by 10 000 sec. illumination at 10 lux. The reciprocity between light quantity and germination is characteristic of the low energy system. When the quantity of light required is small and can be supplied in a single dose, it is conveniently termed a light trigger reaction.

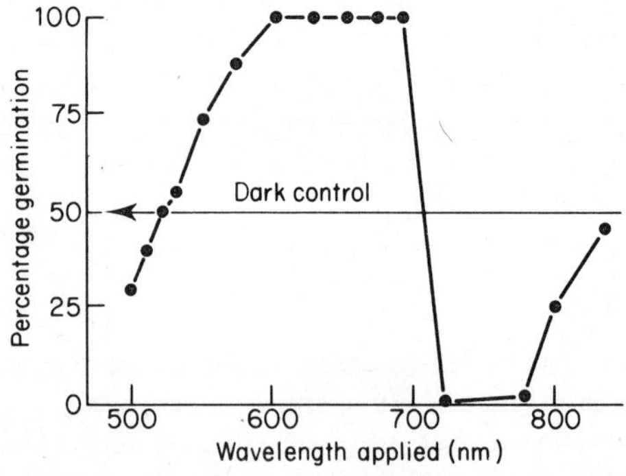

Fig. 3–1 Seeds of Arlington Fancy lettuce which are partly dormant can be stimulated to germinate by red light (600–690 nm) or inhibited by far-red light (720–780 nm). (From FLINT, L. H. and MCALISTER, E. D. (1937). *Smithsonian Misc. Coll.*, **96**, 1.)

Although white light is effective in breaking the dormancy of photosensitive seeds, not all of the constituent wavelengths are equally effective. As early as 1937 it was found possible by the use of monochromatic light to determine an action spectrum for the stimulation (600–690 nm, red light) and inhibition (720–780 nm, far-red light) of dormancy break in *Lactuca sativa* var. Arlington Fancy (Fig. 3–1). This suggests that the perceptive pigment is phytochrome. Light at wavelengths below 550 nm (blue light) was somewhat inhibitory. The effects on stimulation and inhibition by the red and far-red light were entirely reversible and a sequence of treatments with red and far-red light showed that the extent of germination depended only on the nature of the final light treatment given (Table 1).

Table 1 The relationship between percentage germination and the form of final light treatment. (From BORTHWICK, H. A., *et al.* (1952). *Proc. Nat. Acad. Sci. U.S.*, 38, 662.)

Treatment with red (R) and far-red (FR) light				*Percentage germination*
R				70
R	FR			6
R	FR	R		74
R	FR	R	FR	6

The light requirement varies with the duration of imbibition and with the temperature in a complex way that is not well understood. Figure 3–2 indicates the variation in the response of separate batches of seed of a single variety of *Lactuca* to the light or dark at different temperatures. Although the detailed response of each batch is different, germination is only possible within a restricted temperature range.

3.1.2 Light inhibition

Germination of some seeds is inhibited by light. This effect is also controlled by phytochrome. In some species only a brief flash of low intensity far-red illumination is required to inhibit germination; in others treatment with far-red light must be prolonged. For the latter, far-red treatment is equally effective if given either continuously or intermittently, provided the intervening periods of darkness are not too long. Red light given immediately after far-red light will reverse the inhibition.

Although germination of seeds of some species may be stimulated and those of other species may be inhibited by white light, treatment of both types with far-red light converts phytochrome from the P_{FR} form to P_R and thus prevents germination in both. The difference between the two types of response appears to reside in the level of P_{FR} maintained in the seeds in the dark.

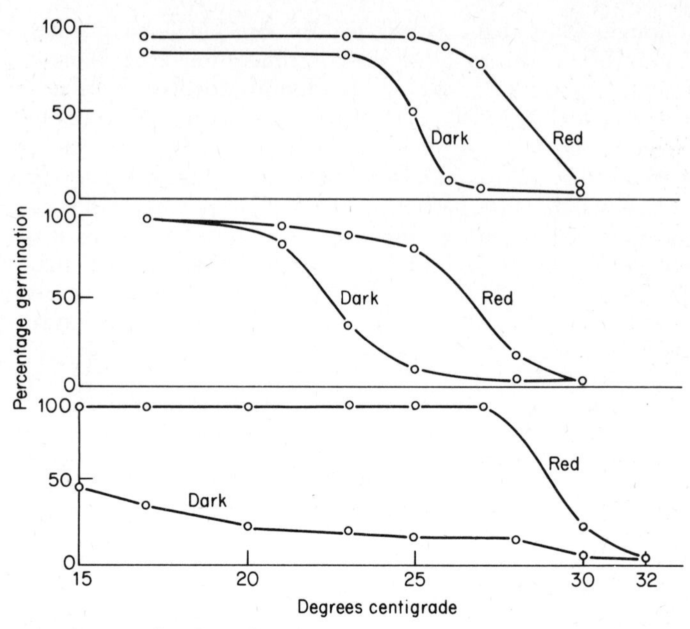

Fig. 3–2 Separate batches of lettuce seeds may respond differently to red light or darkness at different temperatures. (From SMITH, H. (1975). *Phytochrome and Photomorphogenesis*. McGraw-Hill.)

3.1.3 High energy system (photoperiodism)

Some seeds require periodic alternations of light and darkness to promote germination, and the light has to be given over relatively long periods. As in the low energy trigger system the light requirement is quantitative but may vary with temperature. For example at 20°C only a single light period is needed for seeds of *Betula pubescens* (birch) to germinate, whereas at 15°C at least eight consecutive photoperiods are required. Photoperiods of 20–24 h duration are needed for all the seeds to germinate; with shorter photoperiods there is a progressively lower percentage germination. However, most species in this category require only a short photoperiod to induce germination. Longer exposure to light is at first inhibitory, although the inhibition may be released after exposure to a succession of long days (Fig. 3–3). Phytochrome is the photoreceptor. It should be noted, however, that the effect is unlike that involved in flower induction (*q.v.*).

The photoreceptor(s) involved in stimulation or inhibition of germination are not necessarily present in all parts of the seed. Illumination of the cotyledons promotes germination in *Lactuca sativa* but illumination of the radicle is ineffective. On the other hand, light is sensed by the radicle in *Citrullus colocynthis*, where it causes inhibition of

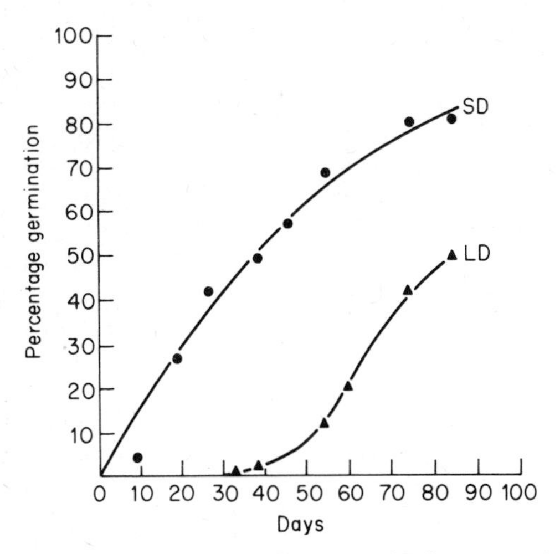

Fig. 3–3 Time course of germination of intact seeds of *Ononis sicula* plants grown in the greenhouse under long (20 h) days (=LD) or short (8 h) days (=SD). (From GUTTERMAN, Y. (1973). In *Seed Ecology* (ed. W. Heydecker). Butterworth.)

germination. Removal of the seed coat has no effect on the inhibition of germination by light in *Agrostis smithii* (bent grass), whereas it abolishes the light inhibition effect in *Cucumis melo* (melon), as well as the light requirement in *Lactuca*. Such varied responses only serve to emphasize the complexity of the system. Moreover, the photoperiodic conditions prevailing on the parent plant at the time of seed development may affect the germination responses of seeds after dormancy.

3.2 Extension growth

Activation of the embryo results in the synthesis of nucleic acids and proteins, in changes of hormone levels, including gibberellins, and in the onset of cell division. Following activation, the seedling uses the stored resources of the seed and begins to grow by extension of root and shoot. To attain photosynthetic independence the two initial requirements for the germinating seed are for the root to extend downwards as an anchor and to begin to make use of inorganic nutrients from the soil, and for the aerial parts of the plant to get above the soil and into the light before the stored nutrients of the seed have been depleted. Early extension growth frequently takes place in the dark when the seed is below ground. Later extension growth of the shoot takes place in the light. The effects of light or darkness on extension growth are poorly understood because several distinct and often apparently contradictory responses are involved; for example light inhibits overall internode elongation, but usually stimulates leaf expansion.

3.2.1 Etiolation

When seedlings are kept for several days in continuous darkness they become etiolated, i.e. the aerial shoots are tall and spindly, the leaves, at least in dicotyledons, do not expand appreciably, and the tissue is pale in colour since chlorophyll fails to develop. Etiolated plants die as soon as the supply of stored nutrients in the seed has been depleted. However, if young etiolated seedlings are put in the light, further extension of the lower part of the shoot is inhibited, the leaves expand, and the plants turn green as chlorophyll is produced (de-etiolation). Photosynthesis can then begin.

Etiolation and de-etiolation represent, in somewhat exaggerated form, the pattern of seedling development under natural conditions. Most of the information on the effects of light on extension growth has been obtained from experiments using etiolated seedlings which have been returned to the light.

3.2.2 Early growth

The downward direction of growth of the root is often influenced by geotropism (a response to gravity) and there is evidence that previously administered light may enhance the geotropic response in some species or inhibit it in others. These responses appear to be phytochrome mediated, but the mode of action is unclear.

Elongation of a shoot results from the continuing formation, by division, of new cells at the shoot apex and the subsequent elongation of these cells. Leaf primordia are initiated at successive nodes on the flanks of the apex. The portions of shoot between successive nodes (the internodes) are at first very short but as their cells continue to divide and expand, so each internode in turn becomes longer, until eventually the extension of a particular internode stops. The overall elongation of the shoot represents the sum of all these growth responses.

3.2.3 Action of light

Both cell division and cell enlargement are commonly stimulated by light. However, as in etiolated seedlings, extension growth may also be stimulated by darkness. How can light bring about these opposing responses? It has been suggested that light acting on phytochrome brings about the activation or repression of key genes associated with protein synthesis. The cellular responses to light are probably indirect. They may result from localized changes in the balance of hormones and metabolites which, in turn, may be due to phytochrome mediated changes in membrane permeability and to the enzymatic changes resulting from the activity of key genes. Each cell undergoes sequential periods of division, differentiation, including elongation, and maturation. One proposal has been that the effect of light on a particular cell would depend on its metabolic state at the time it receives light and causes a rapid switching of the cell from one state of development to the next; a dividing cell would begin to elongate whereas, in an elongating cell, wall development would quickly cease as the cell became mature and elongation would stop. Growth in darkness would, by contrast, delay cell maturation, and, provided a source of metabolites was available (as in young etiolated seedlings), elongation of the shoot would continue unchecked until, on transfer to the light, shoot elongation would stop as the cells almost immediately become mature. However leaf cells of etiolated plants still at the stage of cell division would, on transfer to the light, begin the elongation phase and the leaves would quickly expand.

Such an hypothesis provides a plausible explanation of these apparently diametrically opposed responses in etiolated plants, where light inhibits extension of shoots but promotes expansion of leaves.

3.2.4 Emergence

Initial expansion of the aerial portion of the developing plant is usually confined to only one region of the germinating seedling and it is this part of the plant which on reaching the soil surface first responds to illumination by slowing its rate of expansion growth. Thus in *Phaseolus vulgaris* (French bean) it is the hypocotyl (the region just below the cotyledons) which first expands; in *Pisum sativum* (pea) it is the epicotyl (the internode between cotyledons and first foliage leaves); in *Allium cepa* (onion), the cotyledons; in the *Gramineae* (grasses), the coleoptile. Both blue and far-red light have been found to inhibit hypocotyl elongation (Fig. 3–4) but with blue light there is an immediate inhibition, whereas with far-red light the response is delayed. Two photoreceptors therefore

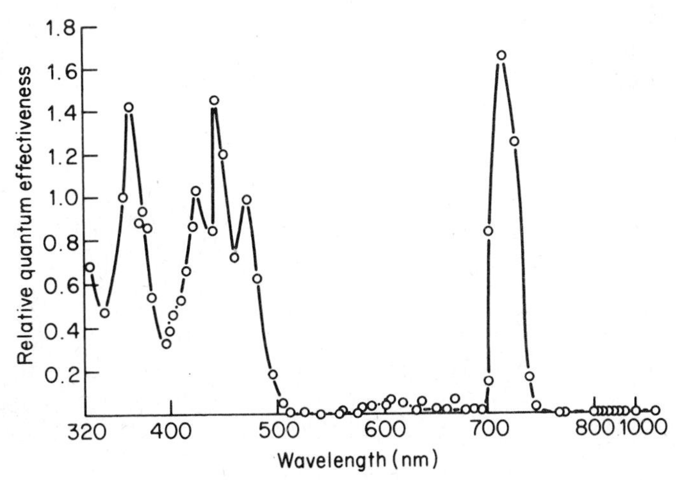

Fig. 3–4 An accurate action spectrum for the high-energy photo-inhibition of lettuce hypocotyl growth. (From HARTMANN, K. M. (1967). *Z. Naturforsch*, **22b**, 1172.)

appear to be involved. The blue receptor has not been identified with certainty; the far-red receptor is phytochrome.

In the *Gramineae*, extension of the tubular coleoptile provides a protective tunnel in the soil through which the young shoot grows. In other species, protection of the young shoot is often accomplished by the formation, in the dark, of a sharp hook within the region of initial elongation (i.e. hypocotyl, epicotyl or cotyledon). It is this hook rather than the delicate shoot apex which, in effect, bulldozes a route for the plant upwards through the soil. After the soil surface has been broken, the hook straightens out. The hook-unbending response is mediated by phytochrome (Fig. 3–5) and is triggered by a single dose of light of low intensity.

3.2.5 *Other growth responses*

Other growth responses to light occur when the seedling has emerged above ground. One such response, which is particularly important in the establishment of effective photosynthesis in the young seedling, is phototropism. This process, described in Chapter I, involves differential growth of the young shoot which becomes reorientated with respect to incident light. The response serves to carry the photosynthetic tissue into the best position to absorb light.

Growth is generally stimulated by light. However, successive leaves and internodes may respond differently. For example, young leaves are importers of metabolites (including hormones) whereas fully expanded

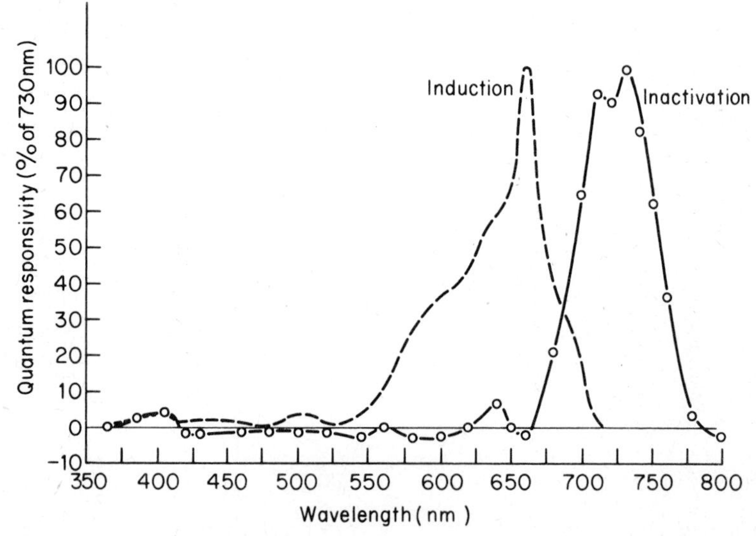

Fig. 3–5 The hypocotyl hook of *Phaseolus vulgaris* unbends on exposure to red light. (From WITHROW, R., B., *et al.* (1975). *Plant Physiol.*, 32, 453.) The unbending response is prevented by far-red light.

leaves are exporters. The expected expansion in light of the youngest leaves may be prevented by the presence of inhibitors produced by older leaves, or by temporary nutrient deficiency caused by the diversion of metabolites to rapidly expanding leaves below. Only when metabolites begin to be exported from these older leaves can expansion of the younger leaves begin. In this way a regular sequence of leaf expansion is established.

Descriptions of growth responses to light are usually based on experiments carried out with monochromatic light which help in identifying the photoreceptors. Under natural conditions the light to which plants are exposed appears white to us, and the value of discovering responses promoted by an artificial stimulus, such as a flash of far-red light, seems at first sight doubtful. However, because the light filtered through an overhead leaf canopy is differentially absorbed, the light reaching seedlings at ground level has an increased proportion of green and far-red light (Fig. 1–8). The light spectrum also changes during each day and at different seasons. Therefore the precise quality of 'white' light at ground level is not constant and the observed responses by plants to specific wavelengths may represent a more significant adaptation to natural conditions than is at first apparent.

3·3 Establishment of photosynthesis

Some lower vascular plants can produce chlorophyll in the dark and so, as soon as the sporeling emerges and is exposed to the light, photosynthesis can begin. In angiosperms, however, chlorophyll is not formed in the dark; only some preparatory developmental processes can take place. Light is required for the completion of chlorophyll synthesis as well as for the organization of the chloroplast lamellar (thylakoid) system and for the synthesis of some chloroplast enzymes. Only then does photosynthesis begin. Each of these photoresponses requires a different light treatment.

3.3.1 Lamellar structure

In meristematic tissue of an angiosperm seedling plastids are present as eoplasts – colourless, spherical or ovoid sacs bounded by a double membrane and containing stroma and sometimes a few fragmented portions of lamellar material (Fig. 3–6 (a)). An early sign of the onset of differentiation may be the encirclement of the plastid by endoplasmic reticulum which is at first smooth but later becomes rough (i.e. ribosomes become attached to it). Starch begins to accumulate; often in sufficient quantity for the grains to distort the plastid (Fig. 3–6 (b)). The starch accumulated at this stage is not a direct product of photosynthesis, but forms following the translocation of reserves stored in the cotyledons. This starch-containing (amyloplast) stage may only persist for a short time before the starch begins to be eroded and the plastid loses its spherical shape to become flexible and amoeboid, often enfolding portions of cytoplasm (Fig. 3–6 (c)). At the same time vesicles appear adjacent to the invaginating inner plastid membrane and perforated stroma lamellae are formed, sometimes in association with rudimentary prolamellar bodies, loosely organized paracrystalline structures. In *Phaseolus vulgaris* all these changes in proplastid structure can take place either in the light or in the dark; furthermore the plastids follow the same sequence of development in all types of cell present in the seedling. In root cells of most species no further plastid development takes place and the plastids continue to function as storage organelles. In the plastids of many cells of the other organs, stroma lamellae undergo further extension, become continuous instead of perforated, align themselves parallel to the long axis of the plastid and soon overlap over short distances to form bi-layered incipient grana or bithylakoids (Fig. 3–6 (d)). These later structural changes are also independent of light in *Phaseolus*, though not in some other species. Although at this stage their thylakoids are structurally similar, the leaves are green in the plants of *Phaseolus* grown in the light but white in those grown in the dark.

Subsequent development of true multi-compartmental grana requires light and the number of grana and depth of stacking vary with light intensity (Fig. 3–6 (g) and Fig. 3–7). In the dark, true grana are not formed. Instead large paracrystalline prolamellar bodies develop, giving rise to plastids called etioplasts (Fig. 3–6 (e)). If plants grown in the dark are returned to the light the prolamellar bodies are transformed (Fig. 3–6 (f)) and the etioplasts soon develop into mature grana-containing chloroplasts (Fig. 3–6 (g)). This conversion of etioplast to chloroplast requires several different responses to light. The initial transformation of the prolamellar body into a convoluted tubular structure has a low energy requirement of less than 10^{-3}J and protochlorophyll is probably the photoreceptor. Red and blue light are effective, but not far-red or green. The rate of transformation of the prolamellar bodies varies with the light intensity over the range 750–75 000 lux. The most effective wavelength is 450 nm, but phytochrome, protochlorophyll, chlorophyll, carotenoids and riboflavin have all been eliminated as photoreceptors. The rate at which grana appear is enhanced by pretreatment with red light and this appears to be partly phytochrome controlled. The rate of chloroplast maturation varies in successive leaves, suggesting some hormonal influence.

3.3.2 *Plastid shape, size and number*

Plastid shape is initially independent of light but light is required for subsequent plastid expansion. Both high intensity red or blue light and low intensity white or green light stimulate chloroplast expansion.

During the periods of cell division and cell expansion in leaves, plastids also divide (Fig. 3–8) and in this way the cell's population of plastids is maintained or increased, producing large numbers of photosynthetically

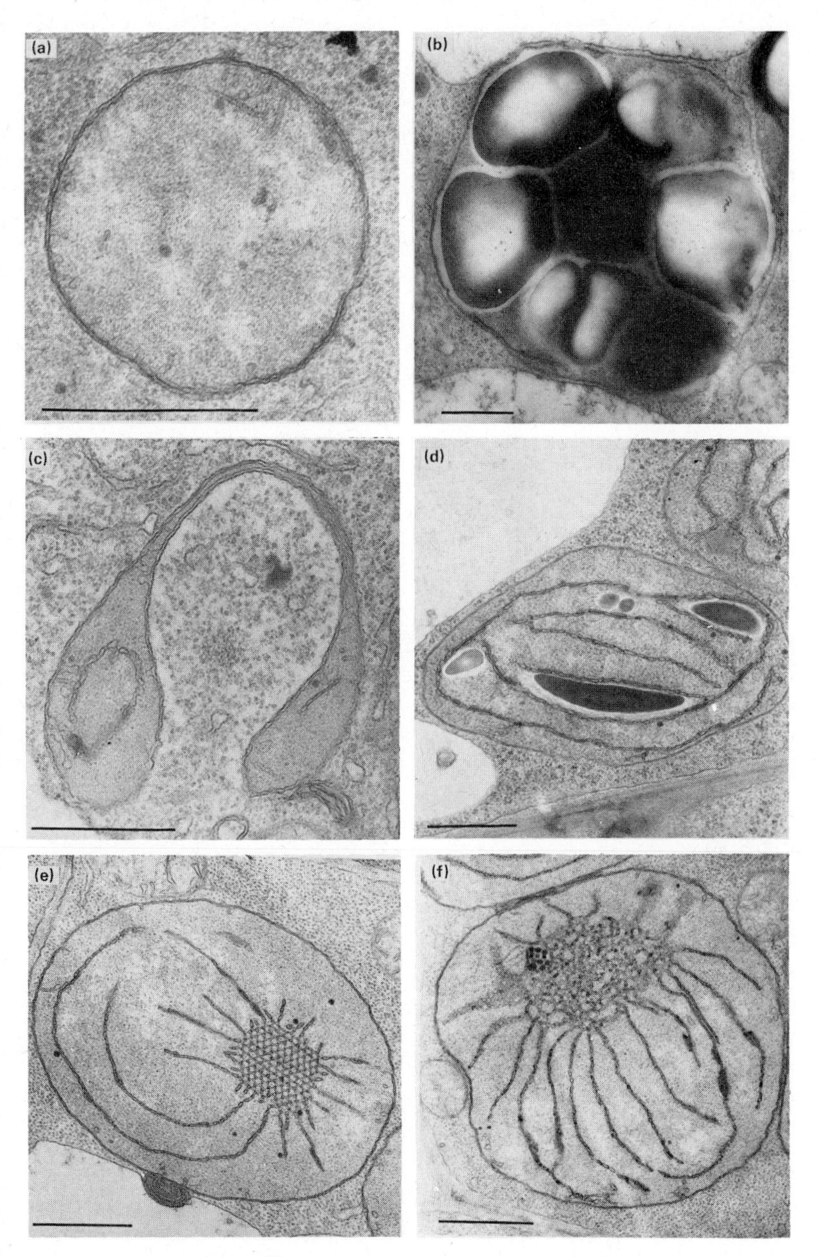

Fig. 3–6 See p. 38 for details.

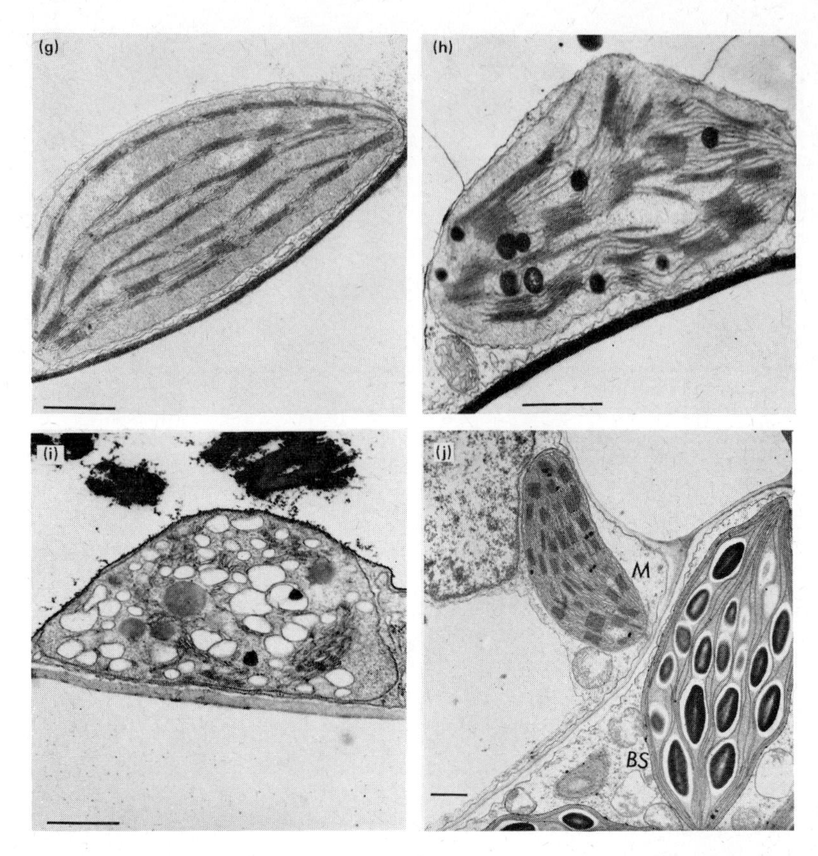

Fig. 3–6 Electron micrographs showing changes in the structure of primary leaf plastids during germination and seedling establishment of *Phaseolus vulgaris*: (**a**) eoplast = 24 h; (**b**) amyloplast = 48 h; (**c**) amoeboid plastid = 3 days; (**d**) pregranal plastid = 5 days; (**e**) etioplast = 5 days growth in continuous darkness; (**f**) etioplast showing early stages of prolamellar body transformation shortly after exposure to light; (**g**) mature chloroplast with grana = 9 days; (**h**) chloroplast showing early signs of senescence = 3 weeks; (**i**) senescent chloroplast = 5 weeks; (**j**) mesophyll (M) and bundle sheath (BS) chloroplasts of the C_4 plant, *Zea mays*. (Scale line = $\frac{1}{2}$ μm.)

active organelles in mesophyll cells of the mature leaf. Plastid division may take place over a wide range of developmental stages from eoplast to recently mature chloroplast. The process involves a preliminary elongation of the plastid as described above. This elongated plastid becomes centrally constricted and finally separates into two daughter plastids of similar size. High intensity red or blue light stimulate division

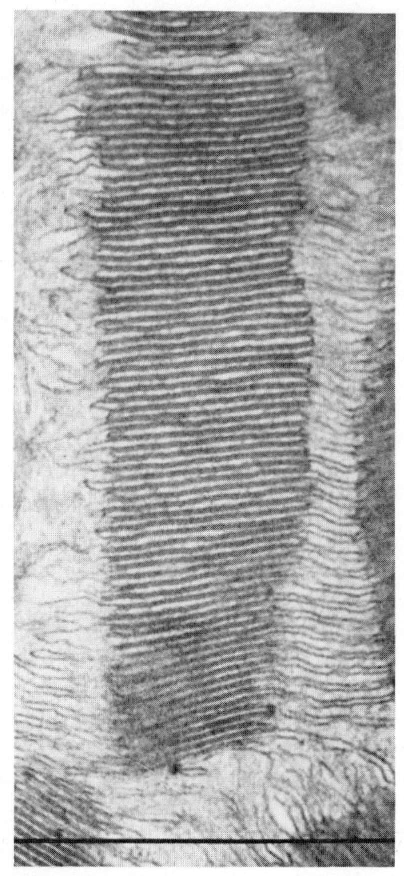

Fig. 3–7 Electron micrograph showing a granum from the shade-plant *Fittonia verschaffelti*. (Scale line=½ μm.)

but low intensity white or green light do not. The photoresponses of plastid elongation and plastid division are therefore different.

3.3.3 Chlorophyll synthesis

The greening of photosynthetic tissues is one of the obvious results that follow the emergence of the young angiosperm plant into the light. If the developing plant, on emerging from the soil, finds itself in darkness (or very heavy shade) it becomes more or less etiolated, the greening process is stopped, material is diverted from its normal uses in development, and, as described above, etioplasts are formed instead of chloroplasts. On placing etiolated tissue in the light the greening process restarts and the chloroplasts finally produced appear quite normal.

The synthesis of chlorophyll itself (summarized in Fig. 3–9) involves the

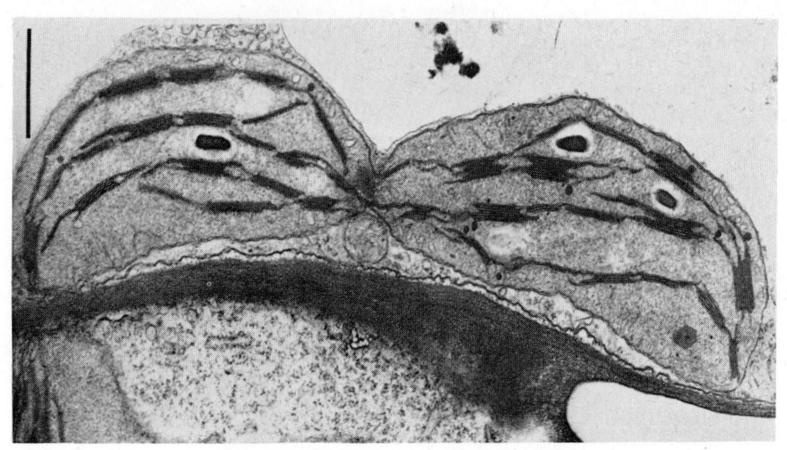

Fig. 3–8 Electron micrograph showing a dividing chloroplast from *Spinacia oleracea*. (Scale line = ½ μm.)

formation of δ-amino laevulinic acid (ALA) from readily available metabolites, the formation of the simple pyrrol, porphobilinogen, and its polymerization and manipulation by decarboxylation and oxidation to yield protoporphyrin IX. These steps are common to the synthesis of all porphyrin compounds, including the haem and chlorophyll series, and take place in the dark. Incorporation of magnesium (Mg) in the centre of the molecule and the cyclization of the propionic acid residue on ring III (also in the dark) leads to the formation of protochlorophyllide, which becomes associated with a specific protein, present in limited quantities in the chloroplast, to yield a protochlorophyllide holochrome. In the absence of light, the synthesis of protochlorophyllide ceases when only a low concentration has been reached and a 'feed back inhibition' appears to prevent the formation of more of the starting compound, ALA. The further metabolism of the protochlorophyllide holochrome requires the photochemical reduction of ring IV to yield the chlorophyll a holochrome. The action spectrum for the formation of chlorophyll a holochrome shows that protochlorophyllide holochrome is the absorbing pigment. The photoconversion is rapid and can be accomplished by a 1 msec flash of intense white light. After conversion to chlorophyllide the long phytyl side-chain is added to the molecule to yield chlorophyll, and the newly formed molecules are incorporated into the thylakoid membranes. This incorporation needs energy which can be supplied by white light and depends on a contribution from the newly developing photosynthetic system.

When chlorophyllide has been converted to chlorophyll the holochrome protein appears now to be able to pick up a further protochlorophyllide molecule, the feedback inhibition of protochlorophyllide synthesis is released and additional molecules of protochlorophyllide

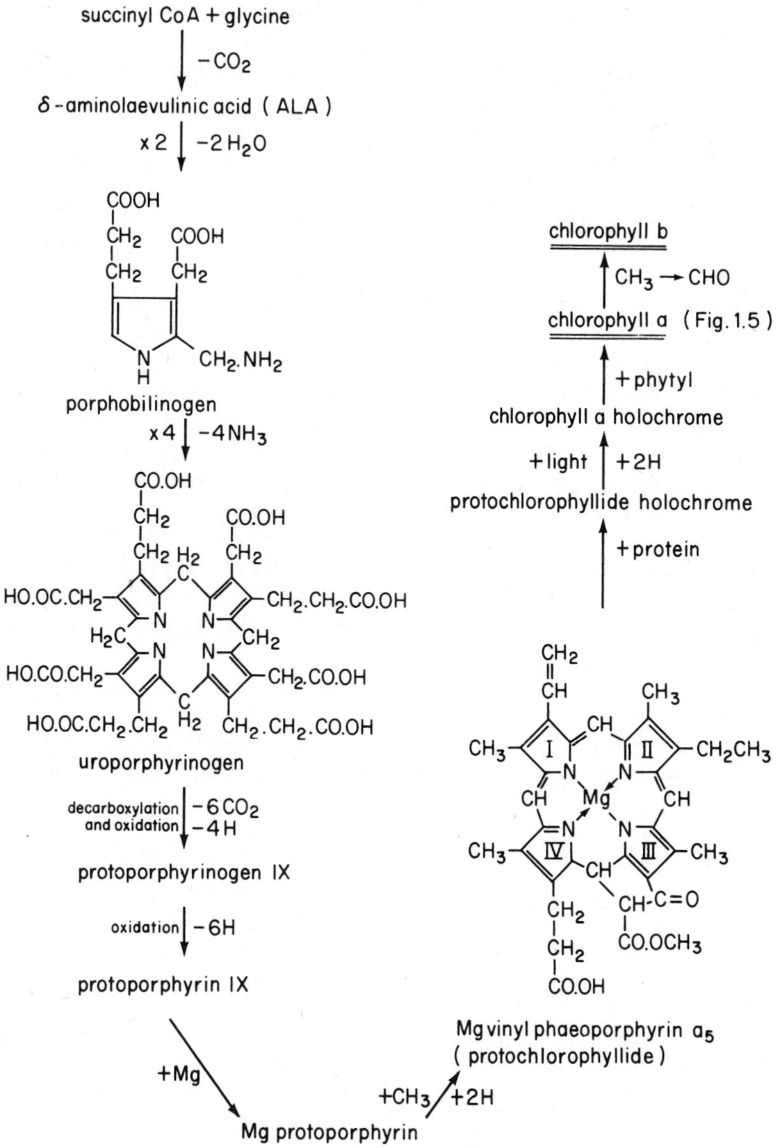

succinyl CoA + glycine

↓ $-CO_2$

δ-aminolaevulinic acid (ALA)

x 2 ↓ $-2H_2O$

porphobilinogen

x 4 ↓ $-4NH_3$

uroporphyrinogen

decarboxylation and oxidation ↓ $-6CO_2$ $-4H$

protoporphyrinogen IX

oxidation ↓ $-6H$

protoporphyrin IX

+Mg ↘

Mg protoporphyrin

↗ $+CH_3$ $+2H$

Mg vinyl phaeoporphyrin a_5 (protochlorophyllide)

↑ +protein

protochlorophyllide holochrome

↑ +light +2H

chlorophyll a holochrome

↑ +phytyl

chlorophyll a (Fig. 1.5)

↑ $CH_3 \rightarrow CHO$

chlorophyll b

Fig. 3–9 Some biochemical steps in chlorophyll synthesis.

holochrome are accumulated. A single flash of light converts all the protochlorophyllide present to chlorophyll, but new synthesis in the dark merely returns the protochlorophyllide to its original low concentration. However in continuous light the photoconversion goes on and synthesis of chlorophylls a and b is eventually limited only by the amount of thylakoid membrane available to incorporate it. Any factor that restricts the formation of thylakoid membranes limits the total amount of chlorophyll synthesized, and leads to the characteristic bleaching (chlorosis) associated either with particular nutrient deficiencies or with growth in the dark.

When etiolated leaves are placed in continous white light there is normally a considerable lag, the length of which depends on the previous treatment of the material, before the major phase of chlorophyll synthesis takes place, even though the conversion of preformed protochlorophyllide holochrome is very fast. However, exposure of the etiolated leaf to a brief irradiation with red light diminishes the usual lag period when the leaf is later exposed to continuous white light; the effect can be largely reversed by giving far-red light immediately after the red treatment. Phytochrome may thus be involved in the control of chlorophyll synthesis and red light may cause an increase in the level of the ALA synthesizing enzyme system previously blocked by feedback inhibition by accumulated protochlorophyllide. A similar effect explains the rapid synthesis of chlorophyll which occurs when etiolated plants of *Phaseolus vulgaris* are given brief flashes of intense white light at intervals of 15 min – a brief flash suffices to remove accumulating protochlorophyll before it can reimpose feedback inhibition, so that the ongoing synthesis of chlorophyll no longer experiences a lag.

3.3.4 DNA, RNA and enzyme protein synthesis

Since all somatic cells in a plant contain the same genetic information the differences which arise between the various cells making up the plant body presumably arise fundamentally from the turning on or off of genes during development. It has been proposed that the primary mode of action of phytochrome results from the activation of certain inactive genes by P_{FR} and the inactivation of other previously active genes. The synthesis of a protein requires coding of the specific amino acid sequence in the DNA of the nuclear gene in the chromosome, the transcription of the code to the nucleotide sequence of the corresponding messenger-RNA, and translation of the message into protein by the cytoplasmic ribosomes under the direction of the messenger-RNA. Use of metabolic inhibitors, known in other systems to affect transcription and translation, have not given definitive evidence for or against the action of P_{FR} on either process, though it seems likely that there is at least an indirect effect of phytochrome at the transcriptional level. If the state of phytochrome alters enzyme levels by activating or repressing key genes which govern metabolism, the resultant secondary effects on the pathway of differentiation brought about in response to the action of light on phytochrome may be very large.

One much studied enzyme under phytochrome control is phenylalanine ammonia lyase (PAL) which produces cinnamic acid and directs metabolism towards anthocyanin and lignin formation.

$$\text{phenylalanine} \xrightarrow{\text{PAL}} \text{cinnamic acid} \longrightarrow \begin{cases} \text{lignins} \\ \text{anthocyanin} \end{cases}$$

The effect of illumination on lignin synthesis can be very dramatic (Fig. 3–10). Formation of lignin in the light coincides with cessation of the extension growth characteristic of etiolation. Once a cell has become lignified no further cell extension can take place. In *Brassica alba* (mustard) seedlings the activity of PAL is very low in the dark. On illumination, after a short lag period the activity increases greatly for up to 24 h, after which it decreases. One interpretation of this result is that PAL synthesis is first induced (perhaps by the activation of a specific gene), that the cinnamic acid and other substances produced then accumulate and inactivate the PAL and finally that the enzyme synthesis becomes repressed. Inhibitor and isotope labelling experiments appear to confirm that protein synthesis is required for the increase of PAL activity and are consistent with a direct control of new enzyme protein synthesis. Other enzymes, unlike PAL, are formed in the dark and their synthesis may be switched off in the light, for example lipoxygenase.

Etiolated *Phaseolus vulgaris* leaves have low levels of most of the enzymes of the Calvin cycle. On illumination, greening and chloroplast development take place and the amounts of the Calvin cycle enzymes increase markedly. Experiments with etiolated plants given brief exposures to red or far-red light clearly indicate that the primary CO_2-fixing enzyme, ribulose-*bis*-phosphate carboxylase, and several other enzymes of the cycle, like triose phosphate dehydrogenase, phosphoglycerate kinase, aldolase, and transketolase, are increased on exposure to red light and that the increase is largely prevented by subsequent exposure to far-red light. These observations indicate a phytochrome-mediated response. Illumination of chloroplasts is also important for the maximum synthesis of triose phosphate dehydrogenase and transketolase, perhaps by supplying essential photosynthetic products needed for their synthesis.

On illumination of etiolated leaves the two photoreactions of photosynthesis become activated and catalytic intermediates, like cytochromes and ferredoxins, are synthesized. Phytochrome (P_{FR}) controls the synthesis of the flavoprotein that catalyses the transfer of electrons from reduced ferredoxin to NADP but not the synthesis of the other proteins of the electron transport system, which simply increase in amount as the chloroplasts develop. It is therefore clear that light acts on several different pigment systems before effective photosynthesis can begin.

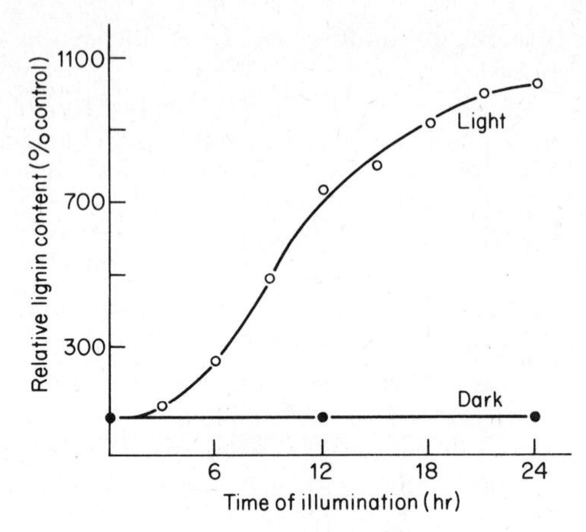

Fig. 3–10 Time course of lignin formation in spinach beet seedlings. (From BUTT, V. S. and WILKINSON, E. (1979). *FEBS* 15, Symp. S8. Pergamon Press.)

3·4 Overall effects on seedling establishment

The overall success of the plant requires its early establishment. When the initial store of nutrients in the seed has been used up, the first leaves must already have become fully expanded and organized to perform their photosynthetic function and to support the development of younger tissues. The effects of a delay in leaf expansion and in the onset of photosynthesis will continue to be felt throughout the life of the plant and may drastically affect its reproductive capacity.

Light influences many phases of growth in seedlings. In *Sinapis alba* (white mustard), in addition to the responses already mentioned, light controls the formation of hairs and stomata, the differentiation of primary leaves, changes in the rates of cell respiration, synthesis of carotenoids and ascorbic acid, and degradation of storage fats and proteins. The various responses to light are integrated in complex sequences of reactions (direct and indirect) and depend on the participation of several different pigments. Protochlorophyll, chlorophyll and phytochrome have been positively identified as photoreceptors; carotenoids or flavins or both may also prove to be so. The precise location of some pigments within the cell is also in doubt, but an organized association with membranes is probable. The responses to light may be immediate or long delayed and the light conditions which affect the parent plant during seed development may in some species determine the photoresponses of its seeds.

4 Further Development of the Plant

4.1 Introduction

At its simplest the shoot is made up of an apical (meristematic) region below which are borne leaves at nodes separated by more or less elongated internodes. The shoot apex itself is protected by small leaf primordia and sometimes by modified leaves, and this forms the terminal bud. Shoot growth results from expansion of the leaf primordia and extension of the short internodes which separate them. During flowering the leaf primordia at the apex are replaced by primordia of the successive floral parts, the sepals, petals, etc., and little internode elongation takes place. Such modifications in development at the shoot apices determine the pattern of growth of the above ground parts of the plant and are often under photoperiodic control. As one might expect the effects of light on root growth are, by contrast, few and indirect.

It is convenient to regard the life cycle of a plant as being made up of a number of overlapping stages or phases, consisting of seed germination, seedling establishment, a period of vegetative growth (juvenile and adult phases), a period of reproductive growth and a period of temporary or permanent cessation of growth (quiescence, dormancy or senescence). Many species undergo these phase changes as a normal consequence of increasing age. For other species, light of a particular photoperiod may either be necessary to trigger the change from one phase of development to the next or it may simply influence the rate of change. The response produced will depend on: (1) the physiological state and overall photosynthetic activity of the plant – the effective response may depend on whether the plant is in a vegetative or reproductive state, and also on the number and type of organs (leaves, tubers, flowers) already present on the plant; (2) the photoperiodic sensitivity of the leaves already present – a leaf is usually most sensitive about the time when expansion ceases, at which time the leaf becomes an exporter rather than an importer of carbohydrates; (3) the particular photoperiodic requirements of the species.

During each phase of development, the basic responses to light are the same as those which we have described for germination and seedling establishment. However, the expression of these responses in the mature plant may be very different from their expression in the seedling. As in the seedling, light conditions perceived at earlier stages of development may also be important.

The seedling is initially dependent upon stored reserves. Later, as photosynthesis becomes established, a net excess of photosynthetic products becomes available. The balance between photosynthesis and respiration changes gradually during the period of vegetative growth and again during flowering and senescence. These changes in the balance of respiration and photosynthesis result partly from changes in the proportion of photosynthetic to non-photosynthetic tissues (e.g. leaves compared with roots), partly from the relative efficiency of the leaf population and partly from definite shifts in developmental pathways such as flower initiation. As during seedling establishment these responses are influenced by other factors such as temperature, as well as by light.

Each phase change which takes place during the life of a plant therefore depends on the physiological state of the plant when an appropriate signal is received, and results in characteristic changes at the shoot apex which influence the next phase of development. A phytochrome-mediated light signal for such a change is commonly perceived by the leaves of plants sensitive to photoperiod and translated there into substances (stimulants or inhibitors) which are then translocated to the shoot apex. Exposure of only a small portion of one leaf may be sufficient. The all-important events which take place within the apex are then responsible for major changes in the form or habit of growth of the plant. Initially, these events include biochemical changes in nucleic acid content and protein synthesis which lead to alterations in the patterns and rates of cell division, and eventually to changes in the gross morphology of the shoot.

In juvenile plants the apex is often quite narrow. Mitotic activity extends some distance down the shoot and the zone of cell elongation which lies below is also extensive. As the plant matures, the apex (at least in dicots) tends to become wider, the depth of the zone of active cell division decreases and the zonation pattern of cells at the apex alters. As the apex changes, the new dimensions of the area given over to the formation of leaf initials may influence the size and shape of the leaves

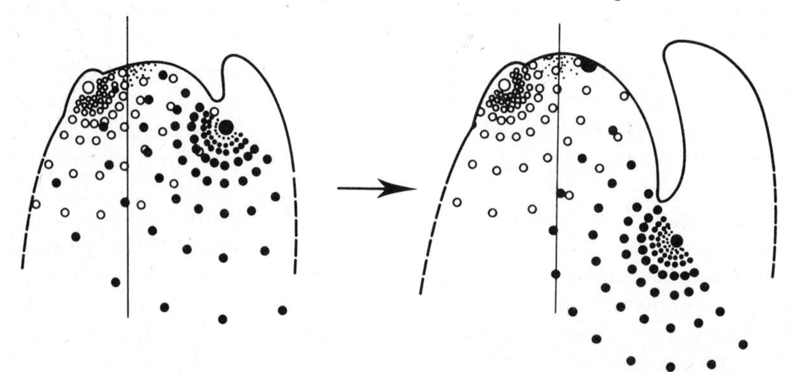

Fig. 4–1 Possible diffusion patterns of hormone, shown diagrammatically, at two stages of shoot development (From SCHWABE, W. W. (1971). *Symp. of S.E.B.*, **25**, 301.)

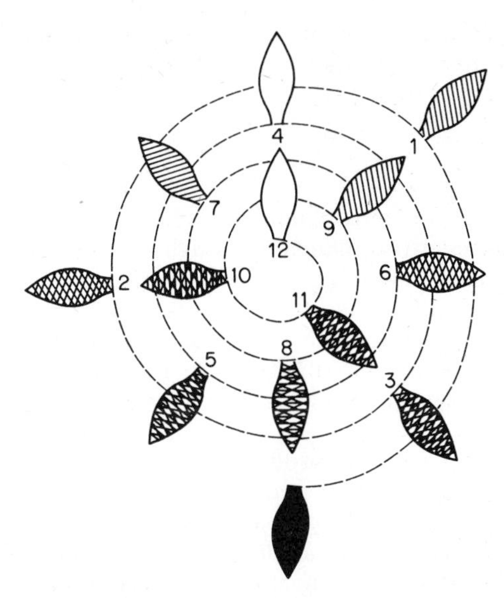

Fig. 4–2 Phyllotaxis and distribution of ^{14}C in tobacco plants. The treated leaf is shown in black. The degree of cross hatching indicates the relative amounts of ^{14}C present in successive leaves. (From SHIROYA, M., *et al.* (1961). *Can. J. Bot.*, 39, 855. Reproduced by permission of the National Research Council of Canada.)

subsequently formed. Similarly the spacing between successive leaf or flower primordia may be altered and this in turn can affect internode length, leaf angle and phyllotaxis. Such morphological changes stem from the creation at particular sites within the apex of new diffusion patterns of metabolites (including hormones) which have been received from a number of sources, but principally from the newly mature leaves (Figs. 4–1 and 4–2). The changing fields of influence of such metabolites result in changing patterns of development of different groups of cells and so eventually help to determine the overall habit of the plant.

4.2 Roots

In addition to its effect on overall growth of shoots, light may also affect growth of other parts of plants, including those parts which lie below ground. Most gardeners know that certain species form rooted cuttings more easily at some stages of growth or at some seasons than at others. The ability to root, the speed of rooting and the number of roots formed may all be influenced by daylength. Night break experiments have shown that these rooting responses are not purely the result of increased photosynthesis during long days. There is a direct response to photoperiod and the leaves are the site of perception. Rooting capacity may be influenced not only by the photoperiod under which a plant grows, but also by the light treatment experienced by the parent or stock plant from which it was derived.

4·3 Leaves

4.3.1 Leaf shape

Leaves are the organs most commonly involved in light perception and are also themselves greatly influenced by light. During the juvenile phase of growth, the leaves are often simple in shape. As the plant matures, leaves of more complex shape are often formed, and when the plant flowers or ages, leaves of yet other shapes and sizes may develop (Fig. 4–3). Each of the sequential modifications in leaf shape may be photoperiodically controlled, the particular daylength response depending on the species.

In succulent species leaves produced under long days, for example, tend to be larger and thinner and to have longer petioles than those formed during short days. In some aquatic species the submerged leaves are very different in appearance from the emergent leaves, and it is perhaps surprising to find that in at least one species the change of shape may again be day length controlled. In *Ranunculus aquatilis* (water buttercup), the finely divided submerged type of leaf can be promoted in response to short days, and the broad emergent form in response to long days.

The leaves of other species show more localized responses to photoperiod. For example the smooth or entire leaf margins of the succulent, *Kalanchoe blossfeldiana*, become serrated under long days and

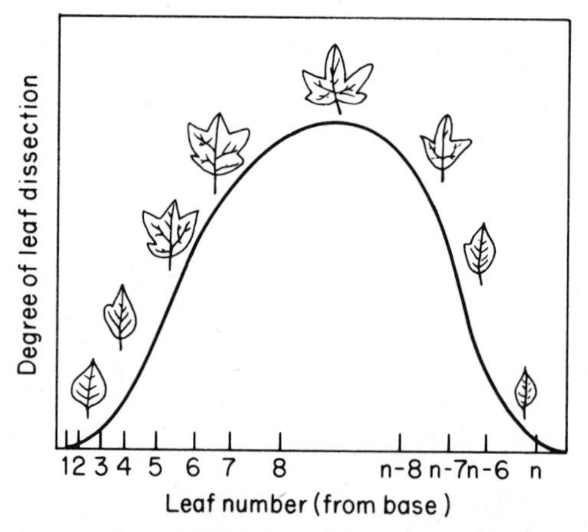

Fig. 4–3 Diagrammatic representation of the sequential changes in leaf form in cotton, from the simple leaf at the lowest node, to the most dissected form on mature wood, to a simple form again near the flowering nodes (represented as *n*). (From ASHBY, E. (1949). *Sci. Am.*, **181**, 22.)

the leaves of *Epilobium hirsutum* become less hairy under short days. The epidermis seems to be an important site of perception of the photoperiodic signal and the upper epidermis to be more sensitive than the lower one.

Changes in leaf shape in response to light may result from various forms of differential growth, for example changes in the area available for the formation of leaf initials at the shoot apex, or from differential growth or development of different types of leaf cell. The extent of vascularization can be influenced by light, and increased development of the main vein, compared with that of mesophyll tissue, can produce leaves which are long and narrow.

The area and thickness of a leaf can be influenced by the duration and the wavelength of the light, but differences in light intensity are responsible for the greatest variations in leaf size. Within limits, the higher the light intensity experienced, the smaller and thicker the leaf, though the precise response depends on the species. Variations in leaf size in response to differences in light intensity are best illustrated by the so-called sun and shade leaves which may be seen growing at different positions on the same plant. The small thick sun leaves mature rapidly and tend to be paler in colour than the larger thinner leaves formed in light shade. However, leaves experiencing deep shade are again small and they mature very slowly.

4.3.2 Internal structure

The internal structure of sun leaves is quite different from that of shade leaves (Fig. 4–4). Sun leaves contain more layers of cells than shade leaves; the palisade cells are closely packed and are elongated in a direction perpendicular to the leaf surface; the intercellular spaces between lower mesophyll cells are small; the vascular system is extensive and cell walls are thick. In shade leaves palisade tissue is poorly developed and

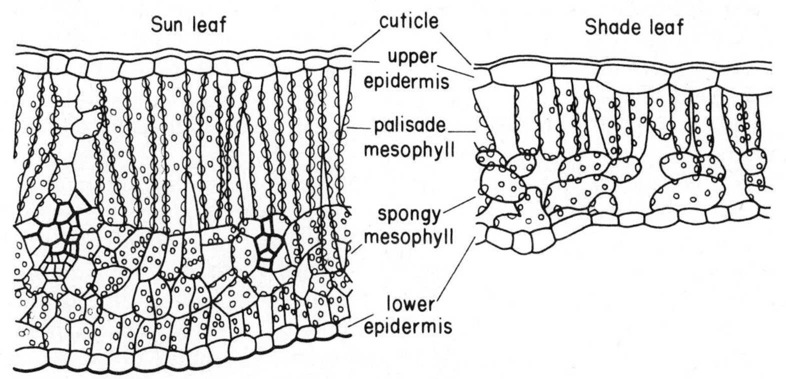

Fig. 4–4 Cross-sections of sun and shade leaves of silver maple (*Acer saccharinum*). (From WILSON, C. L. and LOOMIS, W. E. (1967). *Botany*. Holt, Rinehart and Winston.)

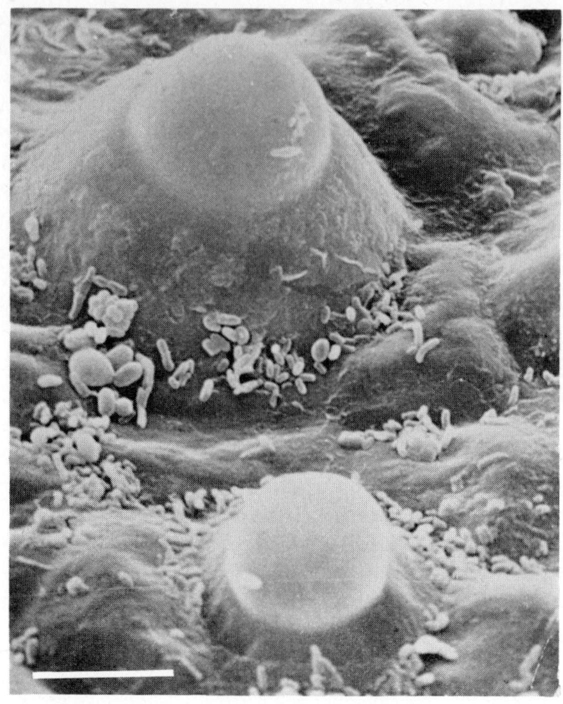

Fig. 4–5 Scanning electron micrograph of ocelli in the upper epidermis of *Fittonia verschaffelti*. (Scale line = 10 μm.)

intercellular spaces are large. The upper epidermis of a sun leaf is covered by a thicker layer of cuticle which is less permeable than that found in shade leaves. Thus excessive water loss by transpiration is avoided in direct sunlight. In a few species of shade plants some upper epidermal cells are modified to form structures such as ocelli (Fig. 4–5). It has been suggested that these ocelli act as lenses and so help to concentrate the small amount of available light on the chloroplasts in the palisade cells below.

It is not only during the early stages of development that the anatomy of a leaf may be modified in response to light, for the leaf is a very plastic organ. As a plant grows the amount of light reaching a particular leaf changes, most commonly as a result of shading by younger leaves, and the internal structure of the leaf becomes adjusted to the new conditions. For example, if a mature leaf is experimentally transferred from shade to sunlight, the isodiametric cells of the upper layers become elongated and so assume the shape of typical palisade cells. This cell expansion results from an increase of soluble sugars at the expense of starch, which in turn increases the osmotic potential of the cells and causes enlargement of the vacuoles. It is characteristic of shade leaves that they have larger chloroplasts with much greater development of granal stacking (see Figs 3–6 (g) and 3–7) than sun leaves of the same plant; there is more chlorophyll per chloroplast, and the ratio of chlorophyll b:a is greater. All these changes result in a more efficient utilization of the less intense and greener light characteristic of shady conditions when sunlight has been filtered through a canopy of leaves. With more chlorophyll per chloroplast the available light harvesting antenna pigments and photochemically active centres are more numerous, and when the chlorophyll b : a ratio is greater the absorption band of the mixed pigments becomes extended towards the green part of the spectrum. If the leaf is exposed to too much light the chlorophyll may become photo-oxidized and the leaf may become bleached. This occurs at a lower light intensity with shade leaves than with sun leaves.

A major function of leaves is to make the ATP needed for photosynthesis. The maximum amount of ATP which can be produced is related to the amount of the membrane-bound enzyme ATP-ase present in the leaf. Recent experiments suggest that on transfer of *Phaseolus vulgaris* leaves from sun to shade conditions the total amount of ATP-ase per leaf remains constant, although the amount of chlorophyll and the extent of granal stacking both increase. In the bean leaf therefore the extent of stacking is dependent on the light intensity, whereas the amount of ATP-ase is not. The rate of photosynthesis per leaf appears to be adjusted to the prevailing light conditions and remains essentially constant.

4.4 Shoot growth: juvenile phase

The juvenile phase of growth is that period during which a plant is not yet capable of flowering. Particularly in some short-lived species flower initials can be formed during seed development on the parent plant or during early stages of germination, in which case the juvenile phase coincides roughly with that of seedling establishment. However, before flowers can be initiated, most species must first undergo a period of vegetative growth, which is often associated with the development of a minimum number of leaves or leaf primordia. In trees this juvenile phase may last a long time, for example in *Betula* (birch) 5–10 years; in *Fagus* (beech) 30–40 years.

The pattern of growth during the juvenile phase is often very different from that during later phases. It is characterized by rapid extension growth, resulting from very great elongation of internodes and sometimes (e.g. in ivy, *Hedera helix*) by an absence of terminal buds. Leaves often differ from those of the adult plant in shape (Fig. 4–3), size, petiole length, phyllotaxis and, particularly in tropical species, in anthocyanin content. In species where flowering is photoperiodically controlled, juvenile leaves are insensitive to daylength changes, and are either unable to produce the stimulus necessary for flowering or unable to transmit this stimulus to the apex. The succulent, *Sempervivum alpinum*, forms plantlets at nodes on runners. These have juvenile leaves which are insensitive to photoperiod and so the plantlets are unable to flower when detached from the parent plant. However the plantlets will flower readily if the runners remain connected to a flowering parent plant. Although these plantlets cannot themselves form a flowering stimulus they can nevertheless translocate the stimulus provided by the parent plants.

Sensitivity of juvenile leaves to photoperiod usually increases at successive nodes as the plant becomes older. The acquisition of photoperiodic sensitivity is an important aspect of the transformation from the juvenile to the adult state.

4.5 Shoot growth: adult phase

Growth during the juvenile phase is more rapid than that during the adult phase, but the growth achieved during both phases is dependent on the amount of light received. Species vary greatly both in their rates of growth and in their tolerance of light conditions, but, in general, established plants grow more slowly under shady than under sunny conditions, and overall growth is usually less under short day conditions than under long days. When plants growing in shade are exposed to light at higher intensities, responses can be both considerable and rapid. In forests, for example, trees of some species (at juvenile or adult phases) may grow very slowly and retain their status as part of the understorey layer for a period of many years. If one of the overshadowing canopy trees falls and a light gap is created, the exposed understorey tree can undergo rapid extension growth and quickly occupy the space made available in the canopy.

Though the increased photosynthetic activity possible following an increase in the intensity or the duration of light is in general proportional to the greater growth achieved, other responses to light may be significant. As explained in the section on seedling establishment, the quality of the light is important, and both red and far-red light may influence extension growth. Stem elongation results both from an increase in the number of nodes formed and from elongation of the internodes. Internode extension, while requiring a continual supply of photosynthetic products, seems also to depend on two phytochrome associated systems, to which the relative response apparently changes as the plant develops. These include: (1) a phytochrome system in the leaves in which exposure to long days promotes internode elongation, the effect being greater when mixtures of red and far-red light are given than when either wavelength is administered alone; and (2) a phytochrome system perceived in the internode itself in which extension is inhibited by darkness, but promoted by far-red light administered immediately preceding the dark period. The spectral distribution of natural daylight filtered through other leaves shows at sunrise and sunset a higher ratio of far-red to red light than at other times of day. Under natural conditions, the additional far-red light which reaches plants at sunset (the critical time immediately prior to the onset of darkness) may therefore be very important to overall plant growth through its effect on internode extension.

4.6 Buds

Woody plants develop as a result of a series of flushes of growth separated by periods of quiescence. In temperate regions many species produce an annual flush of growth followed by a long dormant period in winter. However, in tropical regions particularly, quiescent periods may be so short that growth appears to be continuous. A growth flush involves elongation of the internodes and expansion of the leaves already established at the shoot apex. During quiescent periods shoot expansion stops temporarily. The young leaves at this time may become modified to form protective scales which enclose the shoot apex and the primordia of still younger leaves. These younger leaves will in turn expand during the next flush of growth. Buds can develop at the shoot apices (terminal) or in the axils of leaves (axillary), but the expansion of axillary buds may be inhibited by the hormonal effect of apical dominance.

Several alternative methods of development of the axillary buds are possible in the adult plant. The selection of a particular method is often influenced by photoperiod. Changes which can take place include: (1) death or abortion of the terminal bud; (2) temporary cessation of growth as buds become quiescent or dormant, followed by renewed extension growth of the shoot; (3) permanent cessation of continuous growth of the apex following its transformation into a flower bud (p. 72).

4.6.1 Bud death or abortion

In *Rhus typhina* and *Syringia vulgaris* (lilac) death or abortion of the terminal bud normally occurs. It is daylength controlled in *Rhus* but not in *Syringa*. Death of the terminal bud results in the release of the axillary buds from the effects of apical dominance, and is followed by the ordered expansion of successive axillary buds. This results in a characteristic pattern of growth known as sympodial branching (Fig. 4–6). In species like *Rhus*, daylength control of bud abortion determines the pattern of branching and is thus responsible for the characteristic shape of the plant.

4.6.2 Bud quiescence or dormancy

Many woody species show one flush of growth each year. In spring, buds which are already present expand; new buds are formed and these in turn become quiescent even though environmental conditions seem suitable for continuing growth.

In *Rhus typhina* the single annual sequence of bud expansion and

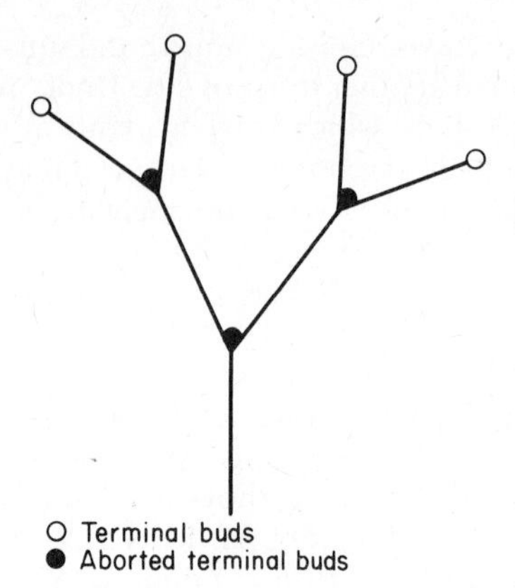

Fig. 4–6 Sympodial branching.

quiescence is under photoperiodic control, whereas in *Pyracantha coccinea* it is not (Fig. 4–7). However, in some species, for example *Quercus* (oak), additional summer flushes of growth may occur (Fig. 4–8). Though the onset and termination of the overall repetitive sequence of successive flushes may be under photoperiodic control, individual flushes are generally in phase with an endogenous rhythm. In *Quercus* this rhythm may be over-ridden and the buds induced to expand prematurely if the leaves already expanded are destroyed by insects. In *Quercus*, and in some other species, the temporary quiescence of these summer buds is easily broken experimentally by a change of daylength. However, the quiescent state of buds often becomes altered towards the end of the growing season. Flushing is no longer easily induced even though environmental conditions remain favourable. These so-called winter buds induced by daylength changes have entered a state of true or deep dormancy.

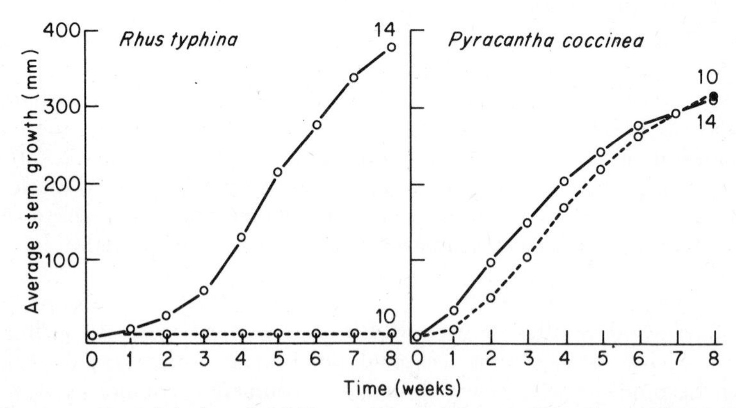

Fig. 4–7 Growth in length of *Rhus typhina* and *Pyracantha coccinea* under 10- or 14-hour photoperiods. (From NITSCH, J. P. (1957). *Proc. Am. Soc. Hort. Sci.*, **70**, 512.)

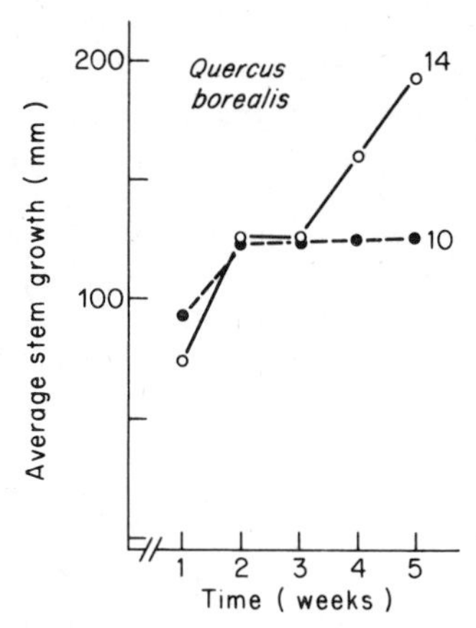

Fig. 4–8 Growth in length of *Quercus borealis* (red oak) under 10- or 14-hour photoperiods. (From NITSCH, J. P. (1957). *Proc. Am. Soc. Hort. Sci.*, **70**, 512.)

The development of such true or deep dormancy requires the photoperiodically controlled development by the plant of a state of cold hardiness, i.e. an ability to withstand very low temperatures. The tissues of many species are normally susceptible to cold injury since freezing may cause mechanical damage by ice crystals or dehydration of the cells. The hardening off process which takes place before winter begins helps the plant to avoid such damage. It includes the conversion of starch to sugars and other small polyhydroxy-compounds – this increases the molarity, and hence lowers the freezing point of the cell sap and helps prepare the plant in advance for the periodically hard conditions of winter.

Subsequent release of the winter buds from deep dormancy may require a daylength trigger (lengthening days of the following spring) in addition to a preliminary stimulus provided by a succession of days of chilling at about 5°C. The period of chilling results in the plant slowly losing its state of cold hardiness and is often experienced by the plant quite a long time before the renewal of growth is triggered by the proper daylength. Without prior chilling, release from dormancy may be long delayed. Though the mature leaves are the most common sites of daylength perception, the buds may also be light sensitive. Without such an alternative site of perception photoperiodic control of bud expansion in deciduous species would be impossible.

Bud dormancy is under the control of phytochrome, as is shown by the reversibility of bud response on treatment with red and far-red light. As for most responses associated with a change from one growth phase to another, bud development is associated with a change in hormone balance.

In warmer regions which experience regular periods of drought (e.g. the Mediterranean area, or dry grasslands) photoperiod is often an important factor in controlling excessive water loss during the dry season.

The shrub, *Sarcopoterium spinosum*, changes its entire habit of growth in response to photoperiod and this allows the plant to adapt to seasonal growing conditions. In the spring season of active growth long branches are formed (extensive internode elongation) with large leaves and spiny apices (a result of increased lignification associated with longer days). These branches are shed during the shorter days of the dry late summer and are replaced by short branches (limited internode elongation) formed in the axils of small leaves. Transpiration is therefore conveniently reduced. During the short days of autumn large leaves are again formed but this time on the short branches. These regular seasonal changes in the pattern of growth allow the plant to grow vigorously when environmental conditions are favourable and to survive when they are unfavourable; each change in habit is controlled by photoperiod.

Under natural conditions most tropical species experience only a slight seasonal variation in daylength; quiescence and expansion of buds are not commonly under photoperiodic control. However some tropical species, when experimentally transferred to the daylength conditions found in temperate latitudes (and therefore unrepresentative of their natural habitat), show photoperiodic growth and quiescence responses which are similar to those of temperate species; a period of quiescence induced in this way is often of brief duration and does not require a photoperiodically controlled release mechanism.

In some tropical areas, however, there are regular periods of seasonal drought. The buds of many plants growing there become quiescent, or perhaps dormant, during this dry season. Further investigation is needed to find out if development and quiescence of these buds are ever under photoperiodic control.

In dry areas of tropical Africa some tree species like the legume, *Brachystegia*, 'anticipate' the onset of the wet season by beginning the spring flush of leaf growth several weeks before the start of the rains. Higher temperatures are believed to be important in producing this response. Although seasonal changes in daylength are small, photoperiod may yet prove to influence the timing of leaf expansion in some species. It has, for example, been shown that at similar latitudes both flowering and seed development in some races of *Sorghum bicolor* are closely controlled by photoperiod in such a way that the time of grain ripening coincides with the normal time of onset of the local dry season.

Different species or races of the same species may have very different requirements for entering and breaking quiescence or dormancy. Indeed, varieties of *Camellia sinensis* (tea) may show continuing growth under tropical conditions and markedly seasonal growth under cooler conditions. Other species may act as perennials in one climate and annuals in another, for example *Phaseolus multiflorus* (runner bean), which is a perennial in its native Central America. Although there is an obvious distinction between the pattern of continuing growth on the one hand and the seasonal onset of true or deep dormancy on the other, a wide range of intermediate states between these two extremes exists and the behaviour of a single species may vary under different photoperiods.

4·7 Vegetative reproduction

Many plants have two quite different mechanisms for reproduction – vegetative and sexual. Vegetative reproduction results from the development of a new plant at a bud on any vegetative organ of a parent plant. The new plant is therefore genetically identical with the parent plant from which it later becomes detached. Sexual reproduction (see p. 59) results from the fusion of haploid gametes formed during flowering. The new plant is genetically different from the parent from which it attains its independence following its development into a seed. Both vegetative and sexual reproduction can be under photoperiodic control.

Vegetative reproduction is particularly common in temperate regions where the same plant may form sexual propagules (seeds) early in the growing season and vegetative propagules late in the season. The grass, *Poa alpina* var. *vivipara*, for example, produces flowers in response to the short days of spring but produces bulbils (vegetative propagules) in response to the long days of late summer. In *Fragaria* × *ananassa* (strawberry) those photoperiods which promote flowering are associated with reduced vegetative growth and, conversely, those which promote vegetative growth are associated with an absence of flowering, i.e. the vegetative growth response is the reciprocal of that for flowering (Fig. 4–9).

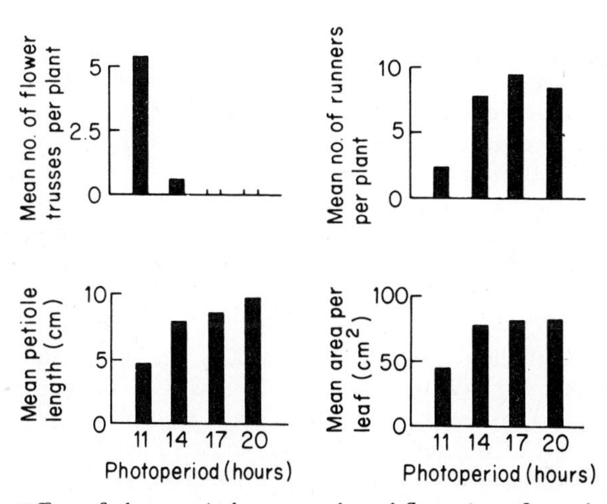

Fig. 4–9 Effect of photoperiod on growth and flowering of strawberry plants. (From BORTHWICK, H. A. and PARKER, M. W. (1952). *13th Int. Hort. Congr.* 801.)

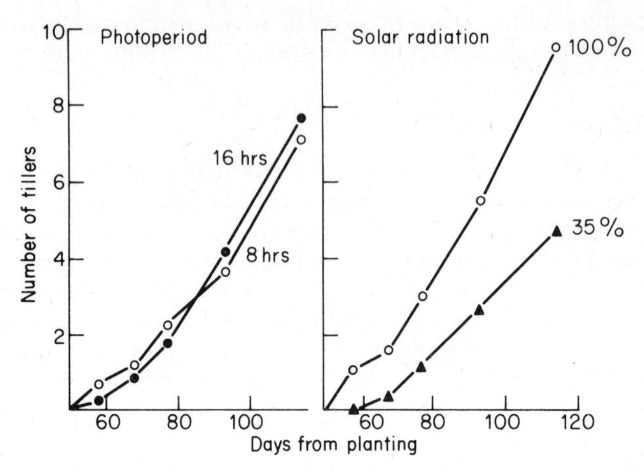

Fig. 4–10 Effects of photoperiod and of level of solar radiation on activation of tiller buds in *Oryzopsis*. (From KIGEL, J. and KOLLER, D. (1970). *J. Exp. Bot.*, **21**, 1003.)

4.7.1 Tillers, stolons, runners and plantlets

There are many different methods of vegetative reproduction. Some species respond to changing daylength by extension growth of axillary buds at or just below the soil surface. In grasses these axillary buds may develop directly into new aerial shoots (tillers). In other species the internodes formed by axillary buds (stolons or runners) may elongate horizontally and new plants may develop at the nodes along these creeping stems. Stolon development in *Fragaria* (Fig. 4–9) occurs in response to long days, but the shoots of *Epilobium hirsutum* behave as runners under short days. By contrast, tillering of the grass, *Oryzopsis miliacea*, shows no daylength response, though the number of tillers formed may increase in response to higher light intensities (Fig. 4–10).

Meristematic cells are present along the edges or at the tips of the leaves of some species of the succulent, *Bryophyllum*. Under long days, these cells differentiate into new plantlets, which eventually become detached from the leaves and are then capable of independent growth. The translocation of an inhibitor prevents plantlet development under short days.

4.7.2 Storage organs

At the end of their growing season many plants produce storage organs, formed by localized swelling of roots (tubers), stems (tubers, rhizomes and corms) or leaves (bulbs). These storage organs represent another method of vegetative reproduction and their formation is the result of several related events. At a time when environmental conditions still appear to favour active growth, photosynthetic products are mobilized and diverted to the developing storage organ and active growth of the shoot stops. Some time after the storage organ has formed it becomes dormant and the rest of the plant dies. With release from dormancy, the adventitious or axillary buds of the storage organ expand and a new plant becomes established at the expense of the stored reserves.

The photosynthetic state of the parent plant influences the size of storage organs but their initiation commonly requires a daylength or a temperature trigger (or both). The leaf is the site of daylength perception, and phytochrome is the sensing pigment (Fig. 4–11). Many tubers and corms appear to be formed in response to short days; the formation of bulbs in several onion species (*Allium*) is in response to long days. In *Allium escalonicum* (shallot) seven days of the correct photoperiod (inductive cycles) are required to initiate bulb development; with a greater number of inductive cycles larger bulbs develop.

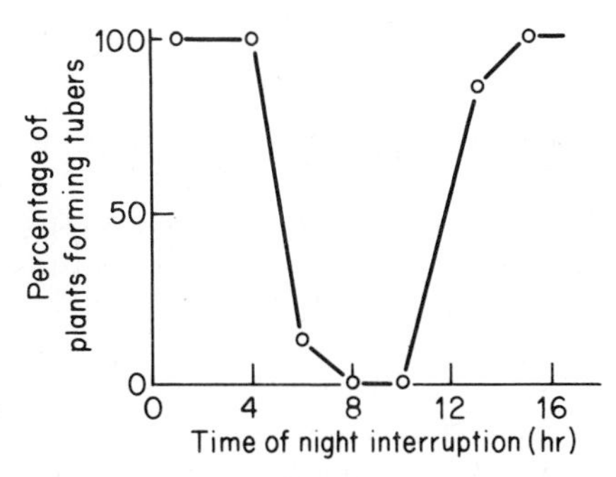

Fig. 4–11 Night interruption treatments can prevent tuber formation in *Begonia*. The light is most effective in the middle of the long dark period. This is a typical phytochrome response. (From ESASHI, Y. and NAGAO, M. (1958). *Sci. Rept. Tōkohu University*, **24**, 81; as drawn by LEOPOLD, A. C. (1964). *Plant Growth and Development*. McGraw-Hill.)

4.8 Flowering

The most thoroughly investigated aspect of the influence of light on changing patterns of growth during the angiosperm life cycle is the effect of photoperiod on flowering. It was realized during the latter part of the last century that daylength might influence flowering, and the importance of a critical period of darkness was recognized soon afterwards. However, it was the famous report by W. W. Garner and H. A. Allard in 1920 that drew world-wide attention to the close relationship between daylength and several plant processes, including flowering. Indeed it did so to such an extent that the fact that flowering often takes place without the benefit of a particular critical daylength is sometimes ignored.

In many species flowering takes place as soon as the plant is old enough, and no additional environmental stimulus is required. In other species no flowers are initiated unless an environmental stimulus (and

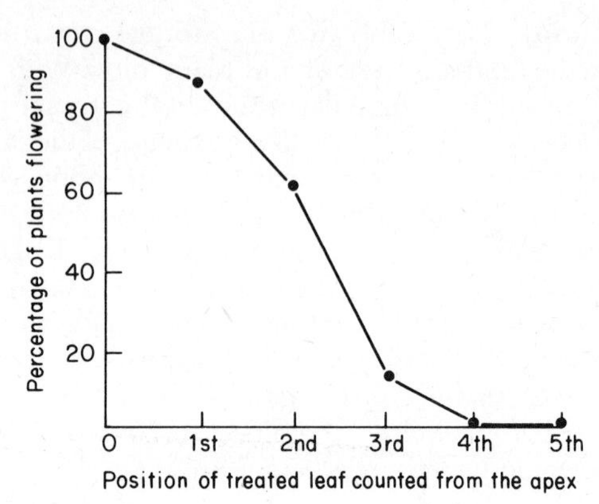

Fig. 4–12 The effect of leaf age and position on flowering in *Xanthium*. The plants were defoliated to leave a single leaf, and this was exposed to one short-day treatment. (From KHUDAIRI, A-K. and HAMNER, K. C. (1954). *Pl. Physiol.,* **29**, 251.)

sometimes more than one) is provided. The most common stimuli are changes in daylength or temperature.

It has already been described how most plants must first undergo a period of vegetative growth (the juvenile phase) before flowering can take place. In photoperiodically controlled species juvenile leaves are at first insensitive to photoperiod, but sensitivity increases at successive nodes and there is a similar progressive change in the capacity of successive nodes to produce flowers (Fig. 4–12). In a few species the provision of a specific number of inductive photoperiodic cycles results in fully developed flowers being formed at all possible nodes. More commonly the number of flowers initiated (Fig. 4–13) or the stage of floral development reached depends on the number of cycles given, and more inductive cycles are required to promote flowering at earlier nodes than at later ones. Under experimental conditions a plant which is capable of flowering when it is old enough without requiring a photoperiodic

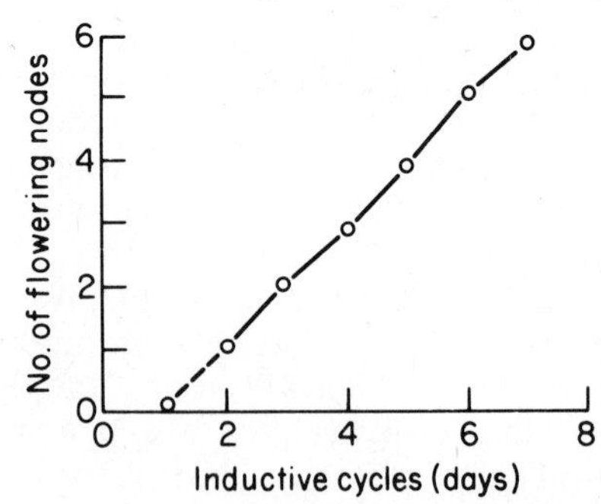

Fig. 4–13 Relationship between number of inductive cycles and number of flowering nodes in soybean. (From HAMNER, K. C. (1940). *Bot. Gaz.*, **101**, 658; as drawn in LEOPOLD, A. C. (1964). *Plant Growth and Development* McGraw-Hill.)

stimulus, can often be made to flower earlier if suitable photoperiodic treatment is provided. When a stimulus promotes such earlier or more profuse flowering the response is called a quantitative response. This differs from the absolute requirement shown by those species which never flower unless an appropriate photoperiod is provided.

Within a local population the simultaneous opening of flowers is important for successful cross pollination. Many tropical species flower simultaneously in response to a brief drop in temperature, and a photoperiodic trigger is not needed. However, other tropical species do respond to photoperiod, in spite of the small annual variation in daylength. Rice (*Oryza sativa*), for example, has been shown to flower in response to a change in the daylength of only five minutes.

Temperate species often show a complex inter-relationship between photoperiod and temperature, and a daylength requirement may be wholly or partly replaced by a response to temperature. Often the precise photoperiod required for flowering depends on the prevailing temperature. However, temperatures experienced during the previous winter may also be important and a period of low temperature is often necessary before a plant will flower the following spring. This is called vernalization and represents a parallel to the chilling requirement needed prior to the release of buds from winter dormancy. Changing daylength during the following growing season then often triggers simultaneous flowering.

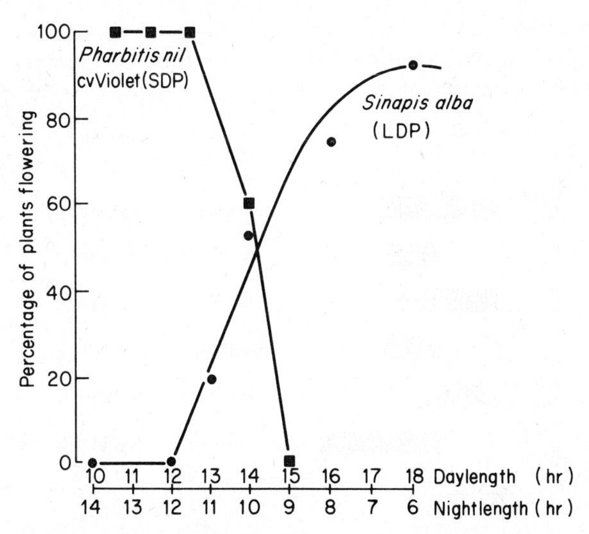

Fig. 4–14 Flowering response of short- and long-day plants to different photoperiods. (Data of IMAMURA, S., *et al.* (1966). *Bot. Mag. Tokyo*, **79**, 714 and BERNIER, G. (1969) in *The Induction of Flowering* (ed. L. T. EVANS. Macmillan, Australia); as drawn by VINCE-PRUE, D., (1975). *Photoperiodism in Plants*. McGraw-Hill.)

4.8.1 Short-day, long-day and day neutral plants

Those species which require either no environmental trigger, or a trigger other than photoperiod in order to flower are called Day Neutral plants: those which require a daylength trigger are called either Short Day or Long Day plants, depending on their particular photoperiodic requirements (Fig. 4–14). As described above, an individual plant may flower in response to a particular photoperiod when it has just reached maturity but be day neutral when it is older.

Periods of both darkness and light are needed to produce a flowering response in photoperiodically controlled plants. The duration, intensity and wavelength of the light are all important. Short-day plants flower only when they are exposed to a sufficiently long period of darkness (this amounts to short days on a 24 h cycle), while long-day plants flower only when they are exposed to a sufficiently short period of darkness (i.e. long days on a 24 h cycle). A long night can be experimentally interrupted by a short period of illumination (30 min or less) and the total dark period may then be 'counted' by the plant as a short night, as shown in Fig. 4–15. The length of the light period is less important and a period of darkness in the middle of a long light period will not cause a short-day plant to flower. Experimental alteration of the daylength to cycles of 18 h or 30 h show the importance of the dark period in determining the flowering behaviour of both groups of plants.

Action spectra for the night interruption effect, showing that red light is effective and far-red is not (Fig. 4–16), indicate the involvement of phytochrome in measuring the duration of the dark period. It is suggested that the slow reversion of P_{FR} to P_R (the *trans-* to *cis-*

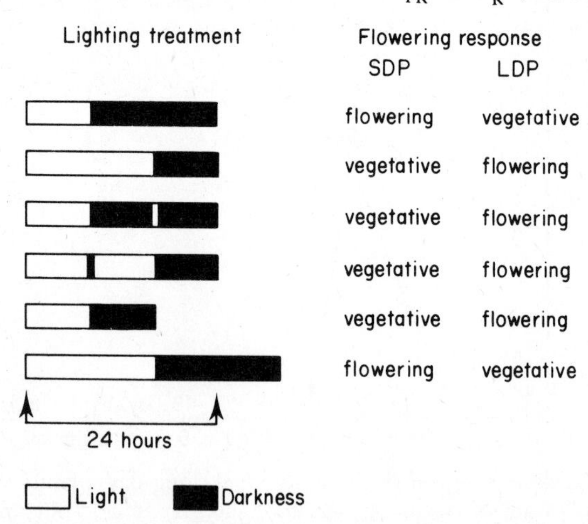

Fig. 4–15 Effect of duration of the dark period on flowering. Short-day plants flower with long dark periods; long-day plants flower with short dark periods. (From VINCE-PRUE, D. (1975). *Photoperiodism in Plants*. McGraw-Hill.)

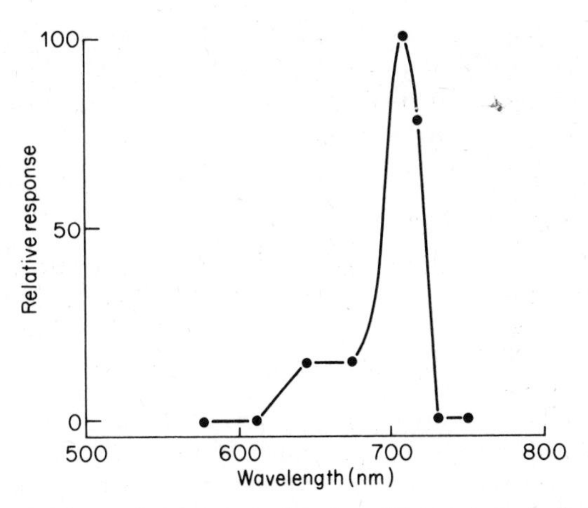

Fig. 4–16 The effects of various wavelengths of light, given during the middle night hours of a daily 16-h dark period, on flowering of the long-day plant, *Hyoscyamus niger*. (From VINCE-PRUE, D. (1975). *Photoperiodism in Plants*. McGraw-Hill. Data of SCHNEIDER, M. J., *et al.*(1967). *Amer. J. Bot.*, **54**, 1241.)

isomerization of Fig. 2–3) which takes place in the dark, acts as an hour-glass timer and that when a sufficiently low ratio of P_{FR}/P_{TOTAL} has been reached in short-day plants the synthesis of the flower-promoting hormone (florigen) is switched on. If hormone synthesis is allowed to proceed for long enough then a flowering threshold is exceeded and flower induction can result. If the dark period is interrupted the light quickly converts P_R back to P_{FR}, thus switching off the hormone synthesis and restarting the hour-glass timing. In long day plants it is necessary to assume that the effect of P_{FR} is to promote flowering by switching off the synthesis of a substance antagonistic to flowering, but this explanation is in many ways unsatisfactory.

In those cases where it has been measured, the reversion of P_{FR} to P_R is temperature sensitive ($Q_{10}=2.0$) in the normal way. However, an important difficulty with the hour-glass hypothesis, which depends on this reversion, is that the ability of plants to measure the night length is to a large extent independent of temperature ($Q_{10}=1.0$). The only biological system known which is essentially temperature-insensitive is the so-called biological clock mechanism that controls diurnal rhythms. Consequently the possible role of endogenous rhythms in measuring the length of the night period has been further investigated. It is proposed that the circadian rhythm is entrained to an exact 24 h cycle by the light to dark transition at dusk or by dark to light at dawn, and that the clock is responsible for setting a biphasic sensitivity to light or darkness, in such a way that interruption of the dark period at a particular time (the inducible phase) may induce or inhibit flowering or it may reset the rhythm. With increasing age of the plant the basic rhythm may become over-riding and flowering may then occur irrespective of the light/dark sequence. The role of phytochrome in this sequence of events is to detect the light signal and to act as a transducer to inhibit or promote flowering depending on the circadian time rather than on sun-time.

4.8.2 Day-length perception

Again, as in other photoperiodically controlled phase changes, the leaves are the usual organs of perception for flowering. The daylength signal is transformed in the leaves into a floral stimulus which is translocated to the apex. The sequence of events from perception to production of the stimulus is called induction; the events which follow the arrival of the stimulus at the apex are called evocation.

Attempts have been made to estimate the rates of movement of the floral stimulus towards the apex after induction. These range from less than 0.1 cm h^{-1} for *Perilla crispa* to 50 cm h^{-1} for *Pharbitis nil* (Japanese morning glory). In *Pharbitis* it took 4–6 h to form the stimulus and transport it to the axillary bud of the induced leaf, but only another two hours to move it a further 100 cm along the stem to another bud. In some species translocation takes place only during the light period and the rate of movement depends on the light intensity. In such species, translocation of the stimulus appears to be associated with translocation of photosynthate in the phloem.

4.8.3 Changes in the shoot apex

As at other phase changes during the life of a plant, flowering involves a change in the structure and the developmental pathway of the shoot apex in response to a change in hormone balance. The hormones present at the apex may well include one or more which tend to promote, and others which tend to inhibit flowering. Alternatively it has been suggested that a single flower-promoting hormone (florigen) may be specially produced. However, at the onset of flowering the balance of hormones appears to swing in favour of that which promotes flowering. The presence or absence of apical dominance helps to determine whether flowering proceeds only at the terminal apex or also at axillary buds. Where apical dominance is strong flowering will take place only at the terminal bud. Where apical dominance is ineffective, axillary buds will expand and the flowering stalk will develop as a branched inflorescence bearing many flowers. The age and physiological (including photosynthetic) status of the plant, as well as other environmental factors, particularly temperature, again greatly influence the precise pattern of development.

When the flowering stimulus reaches the apex, a whole sequence of events appears to be set in motion. During the next two or three days there are increases in nucleic acids and protein synthesis (Fig. 4–17), in numbers of mitochondria and in mitotic index (Fig. 4–18). Each zone of the apex

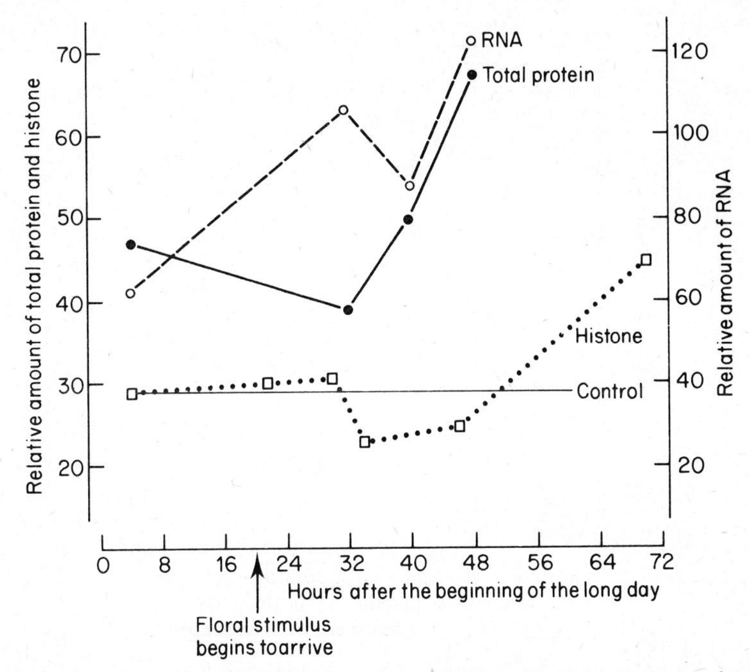

Fig. 4–17 Some changes in metabolic activity following a single inductive long day in *Sinapis alba* (From VINCE-PRUE, D. (1975). *Photoperiodism in Plants.* McGraw-Hill. Data of JACQMARD, A., *et al.* (1972). *Amer. J. Bot.*, 59, 714.)

differs in the rate at which these metabolic events take place. Later the extent of the vascular system below the apex may be altered, for example by a reduction in phloem tissue, and as a result the volume of metabolites transported may change.

The overall effect of the stimulus on the apex is commonly one of increased growth and a doubling of volume has been reported within a period of two days. For example a flat vegetative apex may become domed during flowering. This will alter the surface area given over to the initiation of new primordia, in this case the successive floral parts (bracts, sepals, petals, etc.) which replace the leaves of the vegetative apex. In *Chrysanthemum*, the leaf initials have been estimated to occupy 45% of the apex, whereas the floral bract initials occupy only 12% and floret initials only 1% of the apex.

The preliminary changes in cell metabolism which take place at the apex in response to the arrival there of the flowering stimulus therefore influence the elongating capacity of internodes and the shape, size, structure and phyllotaxis of successive appendages – a pattern of development which parallels that at other phase changes during plant growth.

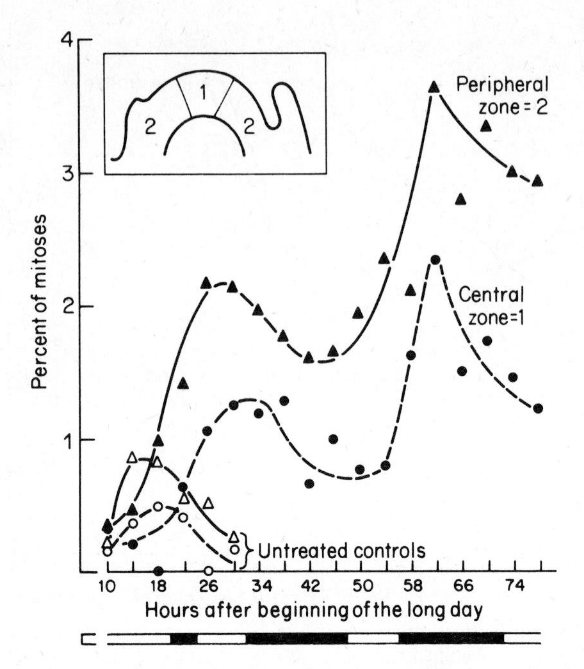

Fig. 4–18 Mitotic index in the central and peripheral zones of the apical meristem in *Sinapis alba* following induction. (From BERNIER, G. (1969). In *The Induction of Flowering* (ed. L. T. EVANS). Macmillan, Australia.)

4.8.4 Inductive cycles

Some species show a 100% flowering response after a single inductive cycle, for example *Xanthium strumarium* (cocklebur); others require many cycles, for example *Chrysanthemum moriflorium* requires a minimum of 12 cycles. In those species which require several inductive cycles to promote full flower development, a smaller number of cycles may, nevertheless, cause a partial response. For example a few cycles may promote elongation of the stalk in plants with a rosette habit, or the formation of the type of leaves (e.g. floral bracts) characteristic of the flowering phase, although the flowers themselves are not initiated. The number of cycles sufficient to cause initiation of floral primorida, may still not be enough to ensure the subsequent development of all the successive floral parts (Fig. 4–19). Further manipulation of photoperiod by the transfer of the plant from long to short days or from short to long days, can in some species even result in the induced floral apex reverting to a vegetative state, with accompanying reversion of the leaves to a vegetative form, or occasionally the formation of leaflike petals or stamens.

Both the photoperiod provided and the number of inductive cycles may influence the sexuality and the fertility of flowers. Some species have

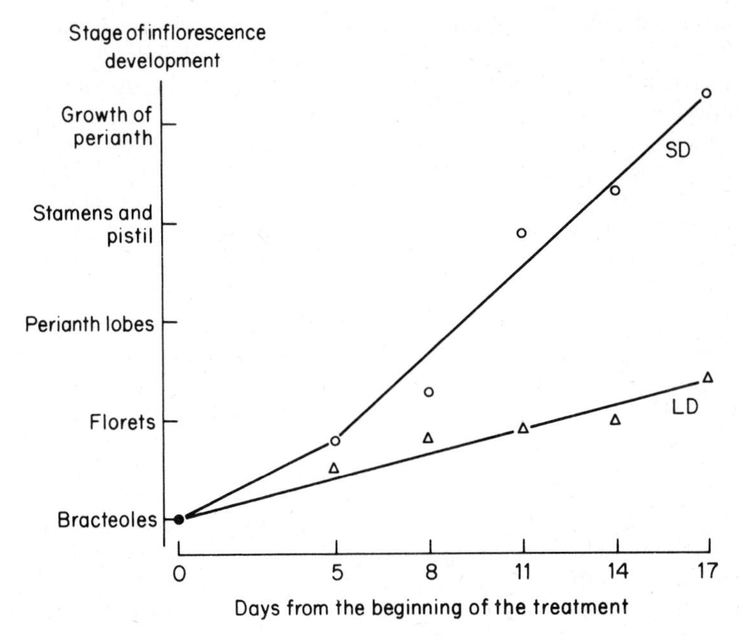

Fig. 4–19 The effects of the number of inductive cycles on the stage of development of the inflorescence in *Bougainvillea*. (From VINCE-PRUE, D. (1975). *Photoperiodism in Plants*. McGraw-Hill. Data of HACKETT, W. P. and SACHS, R. M. (1967). *Proc. Am. Soc. Hort. Sci.*, **90**, 361.)

a flowering requirement not just for long days or short days but for a combination of both. *Xanthium* for example requires short days for flower initiation and development but the provision of long days at a later stage increases pollen fertility.

When too few inductive cycles are provided the flowers or parts of the flowers may degenerate or abort. The tapetum, a nutritive tissue important for pollen development, seems particularly susceptible to degeneration, and when this happens little or no fertile pollen is formed. The precise stage at which degeneration or abortion takes place depends on the number of inductive cycles given.

Species which produce separate and, at maturity, morphologically distinctive male and female plants (dioecious species) can sometimes be induced at an early stage of development to form flowers of the opposite sex. Flowering is promoted in young plants of *Cannabis sativa* (hemp) by exposure to short days. The number of cycles required declines as the plant becomes older. However, the number of short days required to induce flowering in male plants is fewer than in female plants. If short-day treatment is provided early enough, the male plants can be induced to form flowers which are morphologically female.

Photoperiod is not the only way in which light can influence the sexuality of a plant. In *Mercurialis perennis* (dog's mercury) the proportion of male to female plants which survive in a particular habitat depends on the intensity of light experienced rather than on the photoperiod, male plants being more successful at higher intensities.

In species which produce male and female flowers at different positions on the same plant (monoecious species), daylength can influence the proportion of each, and again the precise effect obtained depends upon the number of inductive cycles given. In *Zea mays* (maize), for example, the usually staminate (male) terminal inflorescence can be transformed into a female inflorescence by the provision of additional short-day treatments.

In plants which produce flowers with both stamens and carpels (hermaphrodite species), manipulation of the photoperiod may again alter sexuality and fertility. Stamen or carpel development may be represssed or stamen structure may be modified so that the pollen becomes inaccessible to an insect pollinator. Alternatively, in some self-pollinated species the time of maturation of the stamens and carpels may fail to coincide, and pollination may be prevented. If pollination does take place light may still influence pollen compatibility and so affect fertilization of the egg. Light intensity as well as photoperiod may influence both the earliness of flowering (e.g. in pea, *Pisum sativum*) and sexuality. In *Kalanchoe daigremontianum* light of low intensity reduces or represses stamen development.

4·9 The sporophyte–gametophyte transition

During the later stages of sexual reproduction, as during other phases of growth, a period of dormancy or quiescence may intervene. In some species of the gymnosperm *Pinus* (pine), two years may elapse between the initiation of the pollen and seed cones and the fertilization of the egg. During each of the two intervening winter seasons a prolonged quiescent or dormant period occurs, the first immediately before meiosis and the second between pollination and fertilization.

Periods of quiescence also occur during reproductive stages of the angiosperm life cycle. However, the duration of the process of sexual reproduction in angiosperms is greatly reduced compared with that in other plant groups and resting periods are often brief. Though there is no prolonged quiescent period prior to meiosis as there is in *Pinus*, the onset of meiosis in angiosperms is generally synchronous. Could this result from the use of light as a trigger to end a much reduced quiescent period?

Before the haploid microspores develop into pollen grains there is a quiescent period which may last for a few hours or for several months, depending on the species. A further period of quiescence coincides with the period of pollen dispersal when the ripe pollen grain has become dehydrated.

Many of the waves of cellular change which take place during the course of pollen development resemble those which occur during entry into and release from quiescence or dormancy of other organs. However, the possible effects of light on developmental changes associated with the sporophyte–gametophyte transition have hardly been investigated. It is not known, therefore, if photoperiod is ever responsible for initiating or ending quiescence during pollen or egg development, though in some species pollen germination (extension of the pollen tube) is inhibited by high light intensity.

After fertilization the developing embryo is surrounded by nutritive and protective tissue. These surrounding tissues may be derived from the ovule (the seed) and sometimes also from the ovary and other adjacent tissues (the fruit). Development of all these structures proceeds for some time but then a further period of quiesence often intervenes. During seed development, just as during the development of storage organs, carbohydrates and proteins are diverted from the parent plant. Then, although environmental conditions seem favourable for continuing growth, translocation and synthetic activities cease, and the seed dehydrates and becomes quiescent or dormant. Thus many of the changes which take place prior to seed dormancy again parallel those evident in other organs during the development of cold hardiness and winter dormancy.

The seed or the ripe fruit eventually becomes detached from the parent plant, though the time when this takes place depends on the species. Detachment of the fruit may take place along a specialized abscission zone (see p. 71). Both the onset of seed dormancy and the abscission of fruits are influenced by hormonal changes. In spite of the parallels between seed dormancy and fruit abscission on the one hand and bud dormancy and leaf abscission on the other, there is no evidence of any effect of photoperiod on either the onset of dormancy in the seed or the establishment of an abscission layer associated with the fruit. It is only later during germination of the dormant seed that light once again becomes important in controlling development.

4.10 Senescence and abscission

Senescence is the sequence of changes which take place in plant tissues prior to death. In leaves it involves alterations in chloroplast structure (Fig. 3–6(h) and (i)), chlorophyll content and in the rate of photosynthesis and respiration (Fig. 4–20), an increase in carotenoids and sometimes anthocyanins, and changes in protein and nucleic acid synthesis and in hormone balance. The speed of ageing can be influenced by light. For example, leaves remain green longer under long days than under short days. However, the precise response by leaves to changes in daylength is commonly influenced by the prevailing temperature, and senescence sets in earlier when the temperature is low.

Senescence may affect not only leaves but also other individual organs and whole plants. Some flowers may last for a single day, a plant such as a

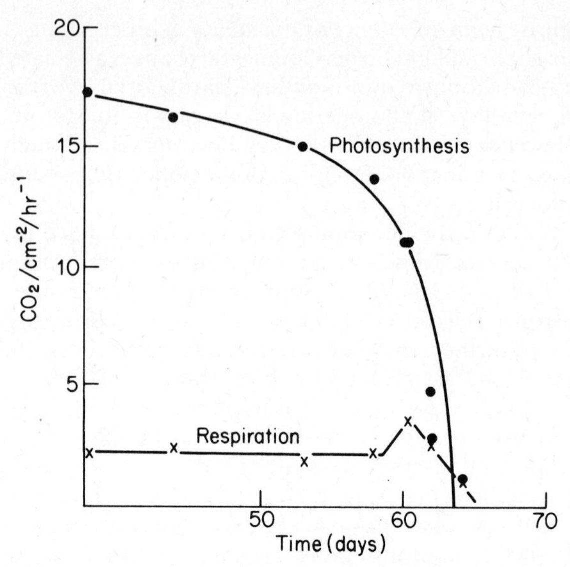

Fig. 4–20 Changes in rates of photosynthesis and respiration of the third pair of leaves of *Perilla frutescens* from completion of leaf expansion to abscission. (From WOOLHOUSE, H. W. (1967). *Symp. of S.E.B.*, **21**, 179.)

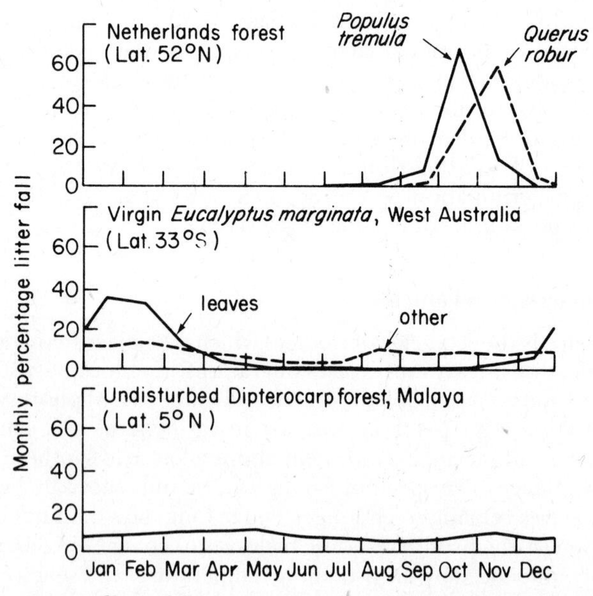

Fig. 4–21 Time of litter fall in some evergreen and deciduous forests at different latitudes. (From BRAY, J. R. and GORHAM, E. (1964). *Advan. Ecol. Res.*, **2**, 101–57. Reproduced with permission. Copyright by Academic Press Inc. (London) Ltd.)

desert annual for a few weeks and a leaf or the above-ground parts of a perennial may last for a season or for several years. Light may influence the life span of whole plants by accelerating the rate of change from one growth phase to the next.

Leaf fall following senescence may require the development within the petiole of an abscission zone. This zone is often formed during early stages of leaf development. Within this zone an abscission or separation layer is established where the cells become separated, and along which the leaves will later become detached from the parent plant. In some species cell division is reactivated and a protective layer of cork may be differentiated on the exposed stump of the petiole. The formation of the separation layer may be under photo-control but abscission is often accelerated by low temperatures.

Evergreen trees remain leafy throughout the year and abscission of senescing leaves can be random. However, for many evergreen as well as for deciduous trees leaf fall is highly synchronized and is often triggered by decreasing daylength (Fig. 4–21), though the synchrony of fall is not so obvious in evergreen species where individual leaves may persist for several years. Abscission of other organs such as flowers or fruits is more commonly brought about by other factors such as low temperature or drought.

Senescence and abscission of leaves in temperate regions and areas of seasonal drought are, like bud dormancy, set under way some time before the onset of winter or the dry season. Leaf fall (like bud dormancy) is important in these regions because it decreases transpiration during a season when water loss is disadvantageous. This is a further example of the predictive importance of daylength as a key feature of plant survival, since sensitivity to photoperiod enables the plant to anticipate future hazardous environmental conditions.

4.11 Summary

Figure 4–22 (p. 72) summarizes the role of photoperiod in controlling the pattern of growth at different phases of the angiosperm life cycle and emphasizes the large number of separate points at which the control may be exerted.

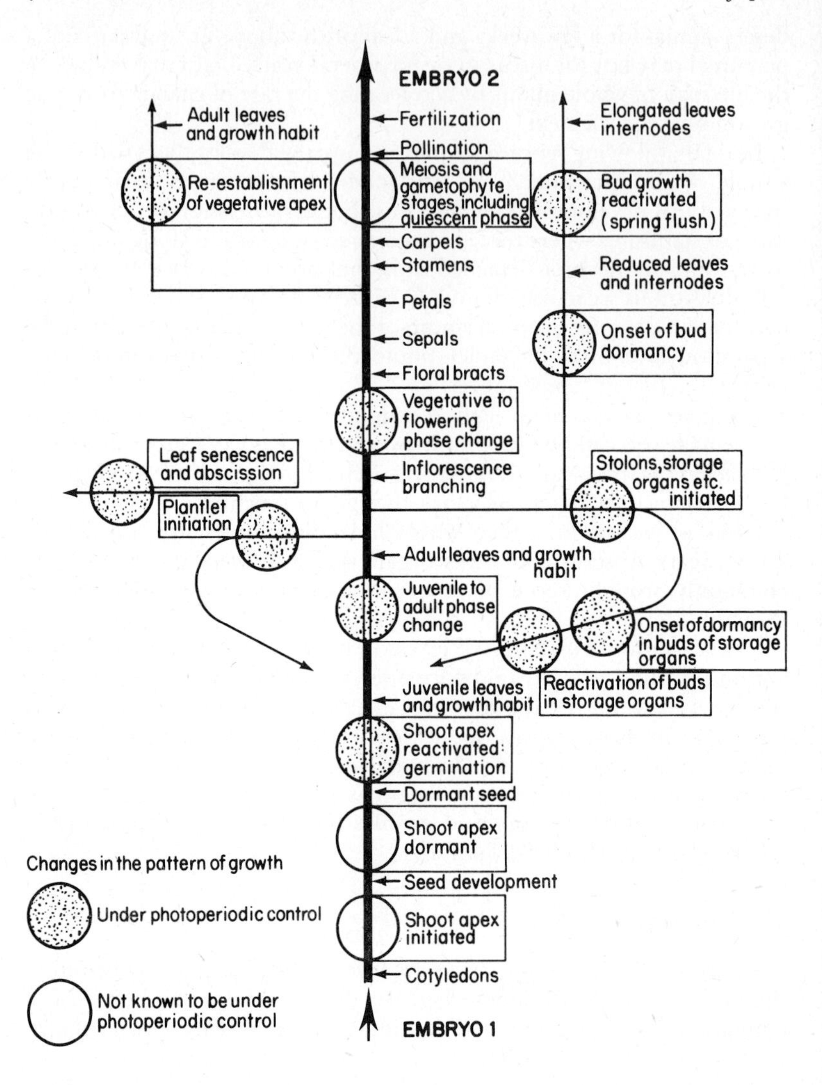

Fig. 4–22 The role of photoperiod in controlling the pattern of growth at different phases of the angiosperm life cycle.

5 Light as a Factor within the Ecosystem

5.1 Introduction

Light acting through a variety of responses plays a major part in the functioning, structure and survival of any ecosystem. The conversion of light energy to chemical energy by means of photosynthesis is the primary process required for the establishment of a food chain. The light requirements of different plant species and the varied habits of growth of these species in response to light result in the development of complex stratified plant communities, which provide a range of habitats for other organisms. Both the exploitation of favourable climatic conditions and the avoidance of unfavourable conditions may be ensured by seasonally timed growth responses to light. It is of major significance that light, as measured by the precession of daylength, is the only completely reliable climatic constant and it is not surprising that both plants and animals make use of this 'clock', although the perception mechanisms evolved are different. In plants the close relationship between daylength perception and those processes directly concerned with population continuity and survival is particularly well marked, for example in reproduction and dormancy. However, although the importance of light as an ecological factor is fully realized, and its role in some processes such as photosynthesis have been investigated over a long period of time, its effect on other ecological processes often remains merely suspected. Most investigations are of necessity limited to single effects on individual species or on simple, usually temperate, ecosystems. The overall effects of light on complex ecosystems remain poorly understood.

The species present within a plant community depend on numerous factors, for example topography, soil, climate (including light), availability of propagules and the structure of the community itself. The structure of the community both affects and is affected by light. For example, in a forest the light which reaches the uppermost leaves of the canopy differs both in intensity (Table 2) and in spectral composition (Fig. 1–8) from that which reaches leaves of plants close to the ground. However, it should be remembered that large differences in light intensity may have very little effect on growth, and it is only when the light intensity reaches some limiting level critical for a particular species that the effect on the plant becomes measurable. For example, for *Impatiens parviflora* overall growth is scarcely affected by light intensities ranging from 20% to 80% of natural summer light, but above and below these levels growth is depressed. This is further reflected in the obvious fact that in habitats where growing conditions are 'good' and the plant community forms a close cover (e.g. forest or grassland) light intensity commonly becomes limiting in the lower layers under the canopy. This limitation is not imposed in habitats like deserts where plants are more widely dispersed and where the low water supply rather than light becomes the critical factor.

Table 2 Minimal percentage day light reaching the ground under different plant species. (Data of (*1*) MONSI, M. and SAEKI, T. (1953). *Jap. J. Bot.*, **14**, 22; (2) IWAKI, H. (1958). *Jap. J. Bot.*, **16**, 210; and (*3*) BLACK, J. N. (1958). *Aust. J. Agric. Res.*, 9, 299.)

Forest (1)		*Herbaceous (1)*		*Cultivated*		
Castanea crenata	13%	*Phragmites communis*	4·5%	Buckwheat aged 37 days	2%	(*2*)
Quercus crispula	7%	*Miscanthus sacchariflorus*	1·2%	Pasture aged 40 days	0%	(*3*)
Field layer below *Quercus*	1·4%	*Helianthus tuberosus*	0·7%			

5·2 Forests

The amount of light which reaches the floor of a forest depends on the species present and on the density of the successive layers below the canopy. Under a dense canopy much of the light which reaches lower leaves is in the form of sunflecks which arrive through gaps in the canopy. When light gaps are small, each sunfleck will persist for a comparatively short period of time, the period depending on the size of the gap, and the changing position of the sun with respect to that gap. Nevertheless, light gaps are an important source of illumination for understorey plants, particularly in dense forests (Fig. 5-1).

Forest communities may be made up of a single layer of trees, or of several layers of trees, shrubs, herbs and grasses. The number of layers present depends on the density of the leaf canopy above. In general, the greater the amount of light which reaches lower levels, the better the development of the lower layers of vegetation.

The lower layers within a forest include not only those species which are permanent inhabitants of these layers and which in non-deciduous forests carry out their whole lifespan under conditions of low light intensity, but also include as temporary inhabitants the seedlings and saplings of potential successors to the members of the upper canopy. Though seedlings of some canopy species are intolerant of low light intensity and cannot survive under the conditions of shade cast by their parents, seedlings of other species are able to do so. For members of the first group, regeneration depends on the seeds falling into clearings which provide better light conditions, or on re-growth from stumps or

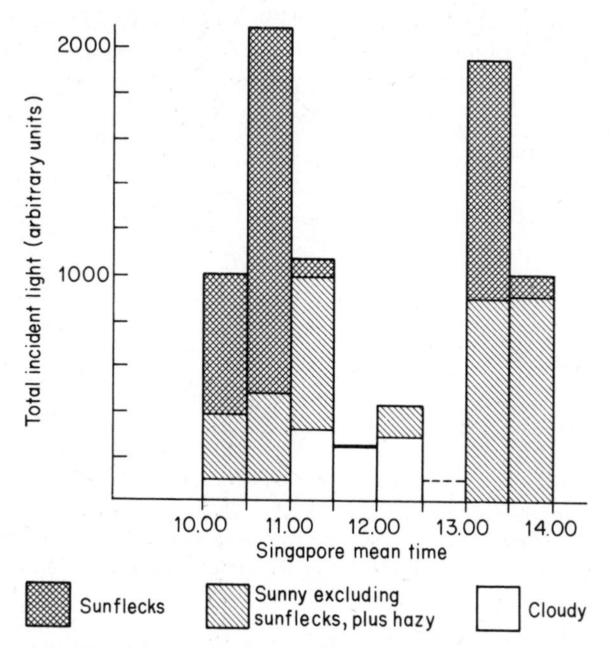

Fig. 5–1 Changes in the composition of mean visible radiation within a tropical rain forest (1° 22′N) during 30-minute periods. (From EVANS, G. C., *et al.* (1960). *J. Ecol.*, **48**, 193.)

suckers. However for those species whose seeds can germinate under low light conditions, subsequent growth of the seedling is often slow and the plant remains dwarfed until a light gap is formed, for example by the falling of a tree close by. The delay may last for many years, but when light does become available, the plant is capable of vigorous extension growth up to the canopy. This light-triggered change in capacity for growth is an important attribute of many forest trees, but the associated physiological changes have been little investigated.

Seedlings of different species vary greatly in their early capacity for growth under different light conditions. However, seeds of species found in dense forests are often large compared with those of species of more open habitats. *Castanea mollissima* (sweet chestnut) with its large seeds and hence large food reserves grows well under shade, whereas *Gleditzia triacanthos* with small seeds fails to do so (Fig 5–2).

Plants present on the floor of a forest must either be capable of growing under conditions of very low light intensity or must be adapted to avoid these conditions. Competition between species for available light has resulted in the development of very different adaptations and the exploitation of very different habitats. In evergreen forests the leaf cover is retained throughout the year, but in deciduous forests there are vast

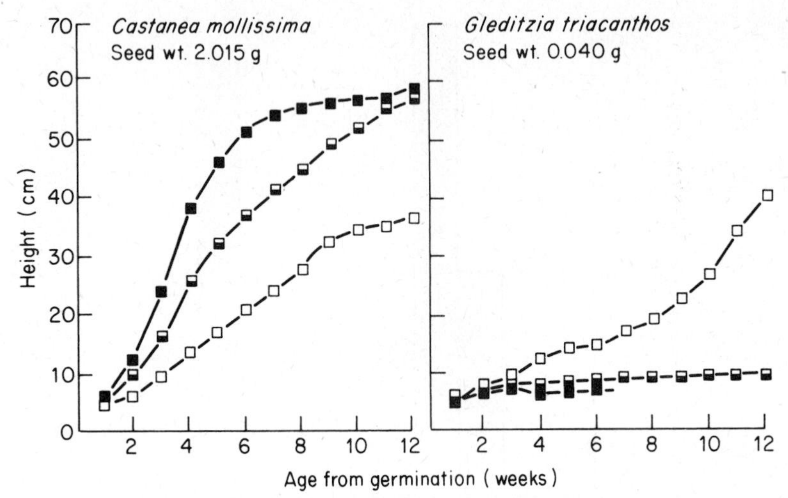

Fig. 5–2 The effect of shading on height growth in seedlings of two tree species. The seedlings were enclosed in 55 cm × 8.5 cm plastic cylinders unshaded (□) and with shade strata of 20 cm (⬓) and 55 cm (■). Height in cm plotted against age in weeks from germination. (From GRIME, J. P. (1966). In *Symp. of Brit. Ecol. Soc.*, **6**, 187.)

seasonal differences in the amount of daylight which can reach the lower levels. It has been realized for a long time that in deciduous forests some ground or field layer species are adapted to these conditions and carry out their entire season's growth before the new leaves of the trees above become fully expanded in the spring. These field layer species die down when conditions become shady (Fig. 5–3). Another type of adaptation

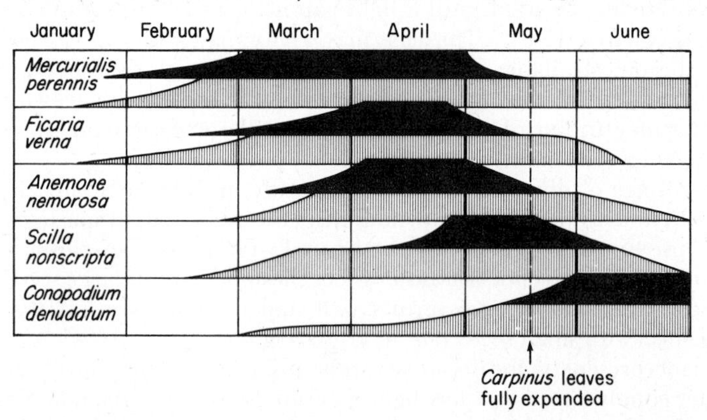

Fig. 5–3 The periods of vegetative growth (cross hatched) and of flowering (black) in the field layer of oak-hornbeam woodlands (*Quercus-Carpinus*) in Hertfordshire, England. (From HARPER, J. L. (1977). *Population Biology of Plants*. Academic Press. Data of SALISBURY, F. J. (1916). *J. Ecol.*, **4**, 83.)

particularly common in tropical forests is the development of a climbing habit. Under the low light conditions below the canopy, such plants produce long stems (cf. etiolation) which continue to extend until they reach the top of the canopy, where light conditions permit active photosynthesis. Other species have become adapted to forest conditions by becoming parasitic and so depend on their host plants rather than on their own photosynthesis for nutrients.

However, competition for light takes place not only between species or individuals of the same species but also between different leaves of the same plant. For example in many dense forests the canopy consists of trees bearing tufts of leaves at the tops of their trunks and very few leafy branches below. Leaves which were present at earlier stages of development at these lower levels have been shaded out as new taller branches were formed, and each canopy tree is entirely dependent for photosynthate on the few leaves remaining on the uppermost branches.

The shapes of leaves and also the angles at which they are inclined play an important part in this competition for light. These differences may be genetic and permanent or environmental and temporary. For example tea plants (*Camellia sinensis*) of different varieties can have leaves permanently set at very different angles to the main stem. Measurements of light below the canopies of these varieties clearly show the influence of leaf angle on light penetration (Fig. 5–4). On the other hand temporary changes in leaf angle may also take place, as they do in seedlings of *Acer rubrum* (red maple). Leaves of *Acer* seedlings growing at high light

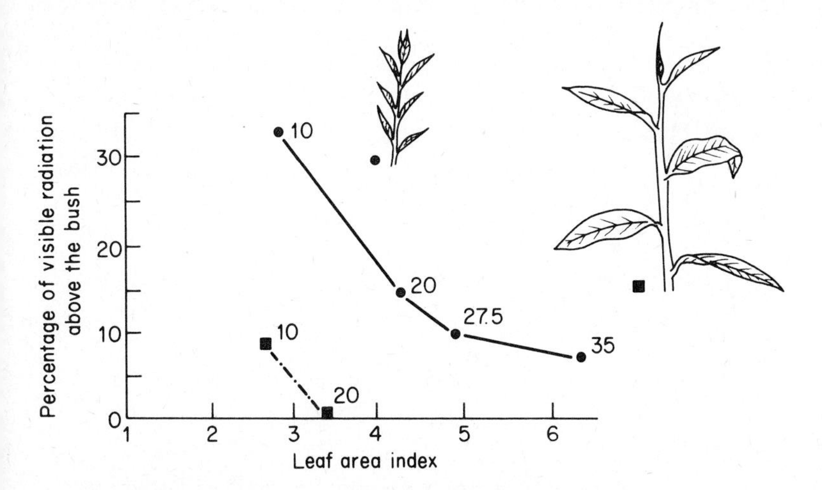

Fig. 5–4 Visible radiation measured on a horizontal surface within the canopies of two types of tea bushes in Assam in relation to the leaf area index above the plane of measurement (depth in cm below the top of the canopy shown beside each point): • ——, near *Camellia sinensis* var. *sinensis*, ■ – • – near *C. sinensis* var. *assamica*. (From HADFIELD, W. (1975). In *Symp. of Brit. Ecol. Soc.*, **16**, 477.)

intensities take up a vertical orientation but later, after shading, the leaves assume a more horizontal position. Competition for light between leaves also results in the development of 'sun' and 'shade' leaves, which are to be found growing at different positions on the same tree (see p. 14).

Let us consider the light conditions within an ash forest (*Fraxinus excelsior*) in the Lathkilldale Nature Reserve, Derbyshire. This forest has ash as its dominant tree species, but the canopy also includes sycamore trees (*Acer pseudoplatanus*).

Where the canopy cover is ash a lot of light passes through and there is a dense field layer, mostly of *Mercurialis perennis*. About 5% of daylight reaches the ground below this field layer and is not used by the *Mercurialis*. However, sycamore trees present a much greater barrier to light and the sycamores on their own allow the transmission of only 5% of daylight. As a result the light intensity under them is too low to support *Mercurialis* and the field layer is therefore very sparse. The ash trees form most of the canopy but the sycamore trees are tending to shade them out. Even within the open ash forest few ash seedlings survive for long (Fig. 5–5). In the darker sites under the sycamore cover any seedlings which germinate increase little in dry weight and are often attacked by fungi which cause 'damping off', a disease to which plants with a low carbohydrate content are particularly susceptible. Where there is more light, ash saplings occupy the middle layer between the *Mercurialis* layer and the canopy. It is impossible to judge the age of the middle layer saplings from their height or diameter alone, but some may be up to 30 years old. These surviving saplings represent seedlings and regrowth from suckers established when

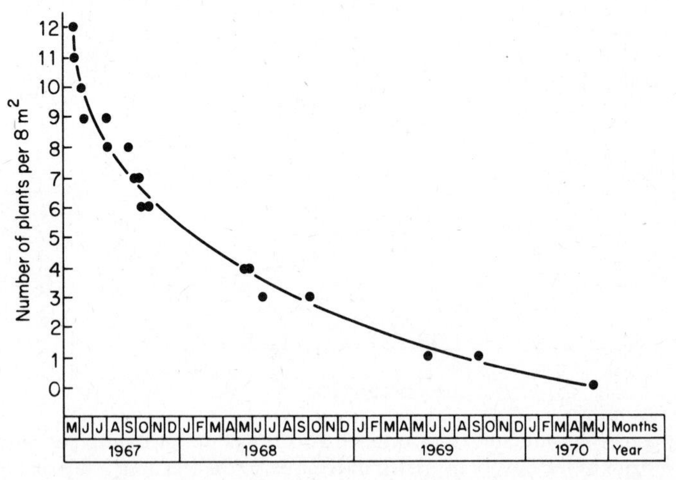

Fig. 5–5 Mortality curve of ash seedlings which germinated in permanent quadrats in Lathkilldale Nature Reserve, Derbyshire, in 1967. (From GARDNER, G. (1975). In *Symp. of Brit. Ecol. Soc.*, **16**, 557.)

light conditions were 'better'. The ash saplings form a reservoir of potential successors to the canopy ash trees (provided, of course, the sycamores do not get there first) and are capable of beginning vigorous growth if lighting conditions again improve, as for example when a canopy tree falls.

5·3 Grasslands, meadows and cultivated fields

The difference in lighting between the inside and the outside of a plant community is best seen by the casual observer in a forest. However, within herbaceous communities or in fields of crop plants, a worm's eye view would show that lighting conditions closely resemble those within a forest; the amount of light reaching the ground again represents a very small percentage of total daylight (Table 2) and comparable changes take place in the spectral quality of that light within the community.

Most grasses are capable of 100% germination in the dark. However, most herbs (as well as most tree species) germinate more readily in the light, even those species normally found in shady sites. Subsequent growth, however, is much slower in shade-tolerant plants than in sun-loving plants or field crops, a characteristic presumably associated with the capacity of young trees in shady habitats to remain as sub-dominants for long periods.

In forests minor variations in topography make relatively little difference to success or otherwise during seedling establishment. However, in more open habitats, minor changes in elevation can have a marked effect on which seedlings survive. Competition for light is important and any seedling which either by virtue of its greater elongation, or by its finding itself on ground raised a few crucial millimetres above its neighbours, can spread its first leaves above those of its neighbours, has a tremendous advantage. This is a reflection of the vertical spread of the community, for what happens within the 20–30 metres of a forest is, in a grassland or meadow community, compressed into less than a metre, even a few centimetres. Competition between layers whether between leaves of the same plant, the same species or different species is otherwise similar in the two types of community.

In an established community it is difficult to determine how important light intensity is compared with other environmental factors during the later stages of a plant's development, and if indeed light, as a single factor, has any important effect at all on the older leaves. However, when fluorescent tubes were placed on reflective surfaces between rows of soya beans (*Glycine max*), it was found that growth was considerably increased in the bottom and in the middle layers of leaves but only slightly in the top layers, which were already fully exposed to sunlight.

In herbaceous communities leaf angle is again important. Variations in leaf angle are associated not only with different species or varieties (e.g. *Camellia sinensis*) or with slight environmental changes, but may also occur

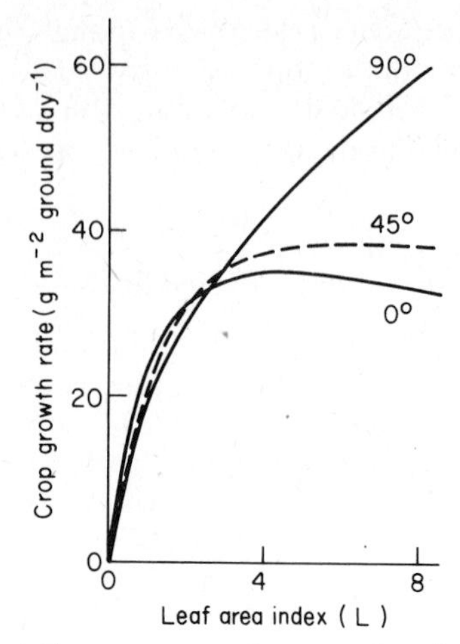

Fig. 5–6 Simulated daily crop growth rates for three hypothetical corn communities, with all leaves at 0°, 45°, or 90° elevation, as a function of leaf area index (L). Solar radiation data for 21 June at Davis, California (38°N latitude), and 10 leaf layers were used. (From LOOMIS, R. S., et al. (1967). In *Harvesting the Sun* (ed. A. SAN PIETRO, *et al.*). Academic Press.)

in successive leaves during the life of a single plant. Growth rates have been modelled for hypothetical communities of maize (*Zea mays*), in which the leaves have been 'arranged' at 0°, 45° or 90° to the horizontal, to try to establish the effect of leaf angle on community photosynthesis and productivity (Fig. 5–6). From such experiments it appears that plants with essentially horizontal leaves (such as dicots) attain their highest growth rates when the leaf area index is less than four (a total leaf area four times as great as the area of ground occupied) whereas plants with essentially vertical leaves (e.g. grasses) do best with an index of greater than four. Some pastures may attain an index of 9–11. Future development of new varieties of 'grass' crops such as maize will probably include an attempt to produce plants with near vertical leaves which can attain a high leaf area index, and which can be expected to give greater productivity per unit area than current varieties.

During the successional stages associated with the establishment of a plant community, there is a change both in the species present and in the density of plant cover, until, at the climax, a more or less stable state is achieved. The habitat is altered, for example by shading, as plants mature; new species with different growth requirements arrive and species already present are eliminated. Each resource, including light, is believed to become increasingly exploited (Fig. 5–7). Plant populations become adjusted both in structure and in growth rate to the prevailing light and other environmental conditions. This can be shown experimentally by sowing plants in plots at different initial densities. The

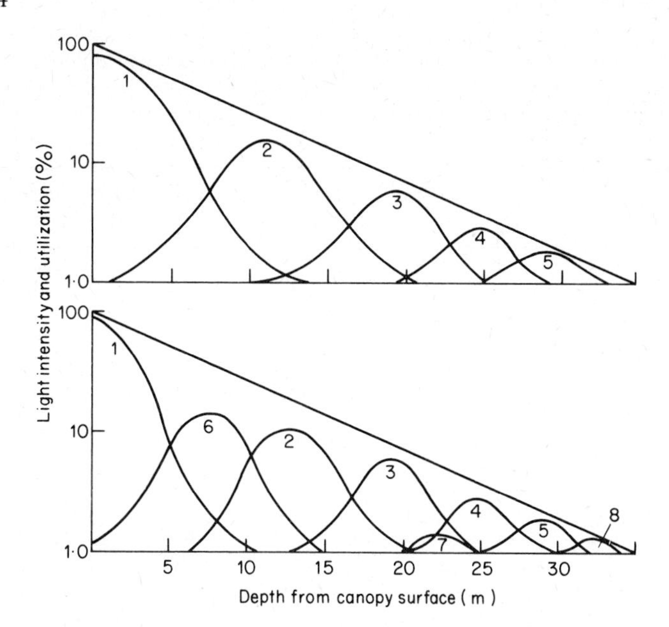

Fig. 5–7 Hypothetical curves showing the adaption over a long period of time of species populations within a forest to different light intensities and the integrated use of the available light. (From WHITTAKER, R. H. (1972). *Taxon*, **21**, 213.)

plants in all plots eventually attain similar densities and light profiles, following proliferation of initially sparse populations and natural thinning out of denser populations. Net productivity also changes from one of excess photosynthetic production as the community develops to a state of balance between photosynthetic input and respiratory loss as the community matures, a change in net production balance similar to that which takes place during the life span of an individual plant.

5·4 Productivity

The total amount of light energy falling annually on the earth is quite large but very little is directly used by plants. In the tropics the intensity of radiation is almost constant all the year round but in higher latitudes there are profound seasonal differences. The amount of cloud cover may also greatly affect the intensity of the radiation reaching the vegetation (Table 3).

Table 3 Annual and seasonal energy inputs (total radiation) in different climatic regions. (From BLACK, J. N. (1956). *Arch Met. Geophys. Bioklim.*, **B7**, 165.) Note that although temperate regions receive low annual input they may receive very large input in midsummer.

	Latitude	*Annual input* (kJ cm^{-2} day^{-1})	*Lowest month* (J cm^{-2} day^{-1})	*Highest month* (J cm^{-2} day^{-1})
Temperate				
Aberystwyth, U.K.	52°N	350	210	1880
Berlin, Germany	53°N	400	170	2010
Sub-tropical				
Davis, California	39°N	670	750	2720
Algiers, Algeria	37°N	690	880	2800
Tropical				
Singapore	1°N	650	1670	2010
Puerto Rico	18°N	670	1550	2130

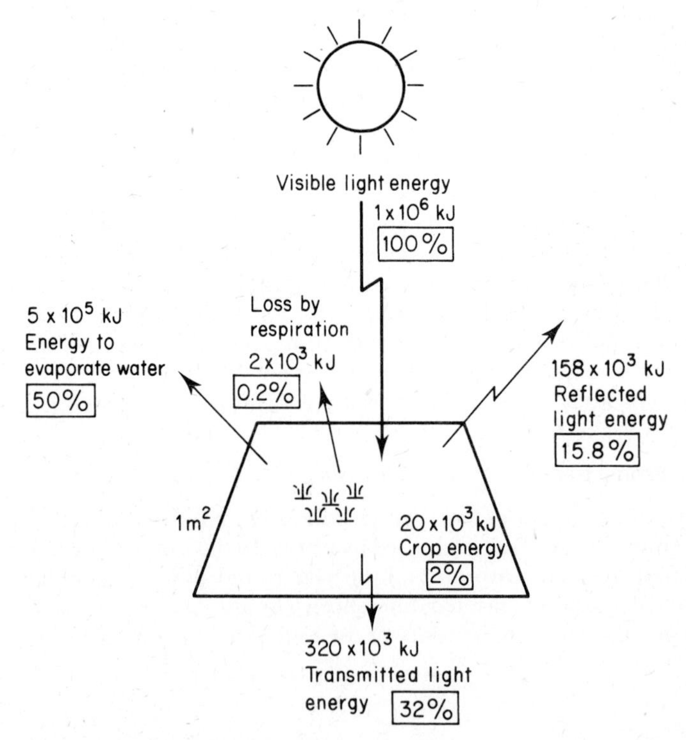

Fig. 5–8 Distribution of visible light energy falling in one year on one square metre of a grassland pasture.

Of the radiation falling on a well-managed pasture (Fig. 5–8), about half is absorbed and converted to heat energy, which supplies the energy for the conversion of liquid water into water vapour. Some of the water evaporated comes directly from the surface of the leaf and the soil and, by affecting cloud cover and humidity, contributes to the effects of climate on the plant. Some of the water loss is by transpiration from the plant itself, and the continued loss of water vapour from the leaf provides the motive force for the transpiration stream, which may be useful to the plant in transporting nutrients in the xylem. Transpiration losses may, of course, be excessive and cause problems of plant survival in areas of lower rainfall. Where combined evaporation and transpiration by the vegetation cover exceed the rainfall over a long period, the active growth of the plants is stopped and each species must resort to its own survival technique (e.g. dormancy) to tide it over to the next growing season.

Approximately one sixth of the radiation is reflected from the surface of the vegetation (albedo) and part is absorbed but re-radiated as fluorescence from the absorbing pigments. Some light may be transmitted through the vegetation cover to the ground below. The amount of light so transmitted varies greatly both with the type of plant cover and with the season.

Only a small percentage of the total incoming light energy is converted into the chemical energy stored as photosynthetic products and new plant material. In the temperate pasture of Fig. 5–8 the net conversion is shown as only 2% and this is much greater than the world average which may be only 0.5%. In a few cases of intensive cultivation the yield, over short periods at least, may be considerably greater than this. Sugar cane for example may show a conversion of 7% at best. For many temperate crops the conversion into economically useful products may increase markedly during seed or tuber growth and at this time the overall conversion of light into chemical energy may double. Under special experimental conditions in the laboratory conversions of 16–20% have been attained in a few cases with algae, but this seems to be the upper limit. It is likely that factors other than light generally limit productivity in natural conditions, for example inorganic nutrients or CO_2 levels (0.03% in the atmosphere).

5.4.1 *C_3 and C_4 plants*

The CO_2 fixation mechanism of a number of tropical and sub-tropical plants, particularly grasses (the C_4 plants) appears to differ from that in most temperate species (C_3 plants). The first detectable fixation product of the C_4 plants is malic or aspartic acid (C_4 acids) rather than the phosphoglyceric acid (C_3 acid) seen in temperate plants. The chloroplasts of C_4 plants commonly show dimorphism, i.e. the bundle sheath chloroplasts contain starch but lack grana, whereas the mesophyll chloroplasts lack starch but contain well developed grana (see Fig. 3–6(j)). Because the C_4 plants, like sugar cane, are more productive than C_3 plants growing in the same environment it was thought that they were more efficient light converters. This is not so. Their biochemical advantage lies in their ability to use the C_4 mechanism to recapture CO_2 that would otherwise be lost by photorespiration. This is particularly important for plants in hot environments where photorespiration can account for a very high percentage loss of newly fixed CO_2, a loss much greater than that experienced by temperate plants. Both C_3 and C_4 plants use the Calvin cycle mechanism for CO_2 reduction but the C_4 strategy requires an additional energy input to pay for the work of running the extra C_4 mechanism for CO_2 conservation. For this reason C_4 plants need more energy, and therefore a higher light intensity, for an equivalent photosynthesis rate than do their C_3 relatives. Paradoxically then, though C_4 plants are more productive because their C_4 mechanism allows them to reduce the loss of CO_2 through photorespiration, they are nevertheless less efficient light converters than C_3 plants. However, over short periods the productivity of some C_3 plants, (e.g. sugar beet) may approach that of the best C_4 plants.

5·5 Light and plant distribution

Plant distribution is primarily influenced by temperature, with water supply as an important secondary factor. Temperature is broadly related to latitude while daylength is precisely determined by latitude. It is therefore not surprising to find a close relationship between plant distribution and the precession of the daylength (see Villiers, Figs 2–2 and 2–4).

Many different processes of plant development are stimulated in response to changes in daylength (see Fig. 4–22). This can ensure plant growth during those seasons when temperatures and water supply are favourable, and plant dormancy or reduction in transpiration area when conditions are not favourable. Attempts to classify vegetation on a regional basis often indirectly recognize this. Thus C. Raunkaier proposed a classification of plants which was based upon their method of surviving unfavourable growing conditions. He was particularly concerned with those structures associated with renewal of growth, i.e. meristems, buds and seeds, and he based his classification on the ways in which these are protected. In other words, this classification is based on those structures which tend to become dormant during unfavourable seasons and are released from dormancy when conditions improve, precisely those structures particularly influenced by photoperiod. The system used by Raunkaier helps to emphasize the very close relationship which commonly exists between photoperiodic control, survival mechanisms and the timing of both active growth and quiescence in regions with regular growing seasons on the one hand and the general absence of photoperiodic control elsewhere.

Though use of daylength as a timing device for growth responses has many obvious advantages, an absolute dependence could have serious disadvantages. The absence from a population of the capacity for variation in response to daylength would limit its potential for long term survival during a period of climatic change or for future extension of its habitat. It has already been mentioned that for many photoperiodically controlled species the requirement of a number of inductive cycles to promote flowering is not an absolute but a quantitative one. For example, a plant which requires short days in order to flower when young may be able to flower without a daylength trigger when older. This ability to over-ride the photoperiodic requirement may be of considerable importance to the survival of the species.

5.5.1 *Ecotypes*

Species of widespread distribution either show little or no response to daylength (e.g. day neutral plants) or are made up of ecotypes each with its own strict photoperiodic requirement. For example, the grass *Bouteloua curtipendula* (side oats) has a wide latitudinal distribution in North America. It achieves this through a number of photoperiodically controlled ecotypes. Strains from Texas are unable to flower under a 15-hour day though they do so readily under their natural photoperiod of 13 hours. Conversely strains from North Dakota, though they flower eventually under a 13-hour day, do much better under the 15-hour day characteristic of their natural habitat (Fig. 5–9).

Ecotypic variation in, or lack of, daylength requirements are likely to

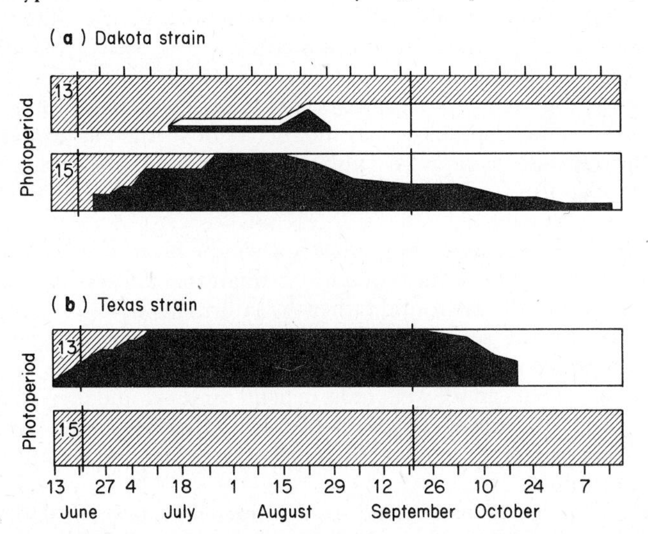

Fig. 5–9 Flowering success of two strains of side oats grown under different photoperiods. (**a**) = Dakota strain, (**b**) = Texas strain; (**a**) adapted to a 15-h day, (**b**) adapted to a 13-h day. For a particular date: black area = flowering; hatched area = percentage not yet flowered; white area = percentage having flowered. (From OLMSTED, C. E. (1945). *Bot. Gaz.*, **106**, 382.)

be of particular significance in relation to species migration in a north–south direction. Few species are found in both the Northern and the Southern Hemisphere, beyond the immediate equatorial region. Although this results from many factors including vast distances and varied terrain, one should nevertheless consider the complexity of the changes required for the spread of, say, a long-day flowering controlled species from an area with a May to September growing period in the Northern Hemisphere, across an equatorial region of minimal daylength variation, to an area with a November to March growing period in the Southern Hemisphere.

5.5.2 *Pollination*

Not only does photoperiodic control enable a plant to time its growth cycle to coincide with the season when conditions are most favourable, but precise photoperiodic control of flowering ensures the availability for cross pollination of a simultaneous and sexually balanced supply of fertile pollen and eggs, an important factor in species survival.

Light may also influence pollination and hence population survival in other ways. Many angiosperms depend for cross pollination on insects, birds or bats which in return receive food from the plant, for example nectar. The selection, evolution and distribution of such plants is closely linked with the lives of their animal associates. Successful pollination often depends on the ability of the flower to attract the pollinator. It usually does this by stimulating the animal's sense of smell or vision, and so 'guiding' it to a source of nectar.

The attracting pigments (e.g. anthocyanins and carotenoids) present in the petals or other floral parts may themselves be formed in response to light. The response by the animal also depends on light. The evolution by the plant of appropriate flower colours attractive to specific pollinators is of selective advantage.

Bats generally feed at dusk and that is when the flowers on which they feed are phased to open. The bats are non-migratory and so depend on a supply of flowers all year round rather than during a limited season, i.e. on a succession of seasonally flowering species or on continuously flowering species which are not under photoperiodic control. Though some species of bat can see well, most depend on sonar and their sense of smell to find the flowers. Flowers of bat-pollinated species are not therefore dependent upon visual identification and tend to be dull white or green in colour.

Birds, on the other hand, tend to be attracted by bright colours. In flowers which are bird-pollinated bright colours often occur in distinctive combinations, such as the orange and purple of bird-of-paradise flowers (*Strelitzia reginae*). Pigment development within the petals is often not uniform but includes streaks of sharply contrasting colours which 'guide' the birds towards the nectar and so towards the pollen.

Insects are by far the most common animal pollinators and many have colour vision, for example butterflies and moths, bees, flies and some beetles. In addition to being able to recognize and be attracted by flower shapes which resemble possible sexual partners, for example bees and bee orchids, insects are also attracted to flowers of different colours. Many insect species have colour preferences, and so selectively pollinate particular types of flowers. Among butterflies, tortoiseshells prefer yellow, and, to a lesser extent, blue, whereas the brimstones and swallowtails prefer blue and purple. Moths which feed at dusk tend to visit only certain white flowers. It is suggested that the white of these flowers is particularly conspicuous when light intensity is low.

Bees are the insect pollinators which have been most closely studied. In addition to their use of polarized light for orientation of flight towards their source of nectar, bees make use of a spectrum of visible light which differs from that of man. Bees show greater sensitivity than man to shorter wavelengths (e.g. ultraviolet) and less sensitivity to longer wavelengths (e.g. red) (see Fig. 5–10). As a result the flower colours distinguished by the bee are different from those seen by man. For example, to a bee the 'white' petals of *Prunus avium* (cherry) presumably appear what we would call 'blue-green' and the 'yellow' flowers of *Potentilla reptans* would appear 'bee-purple'.

Many insect pollinated species (an estimated 30%) have flowers with differentially coloured nectar guides visible to man. However, it is suggested that almost as many species again have guides which can be

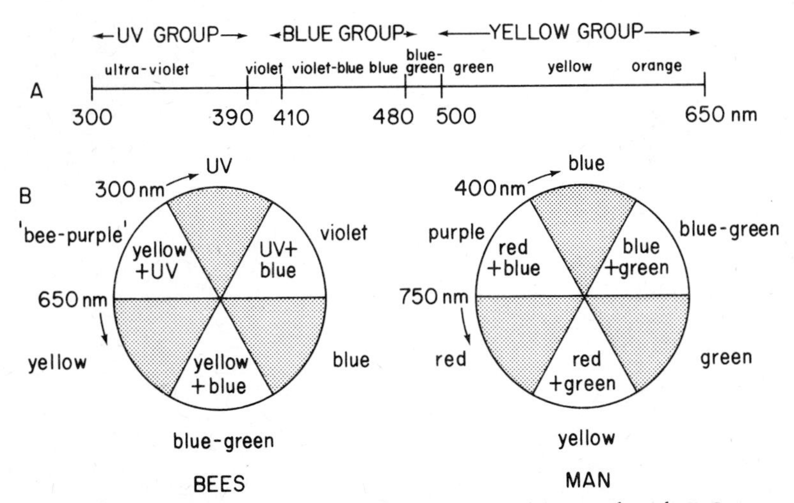

Fig. 5–10 **A**, Spectrum of bee-visible light. Figures show wavelength. **B**, Systems of complementary colours for bees and man. Opposite segments are complementary and when combined produce 'white'. Main colour groups shaded. (From PROCTOR, M. and YEO, P. (1973). *The Pollination of Flowers. New Nat. Ser.*, **54**, 171. Collins. Data of DAUMER, K. (1956). *Z. Vergl. Physiol.*, 38, 413.)

detected by man only in ultraviolet light but which would be obvious to insects like bees (Fig. 5–11).

The effects of light on pollination biology are therefore highly diverse and of considerable importance in the reproduction and hence the selection and distribution of many plant species.

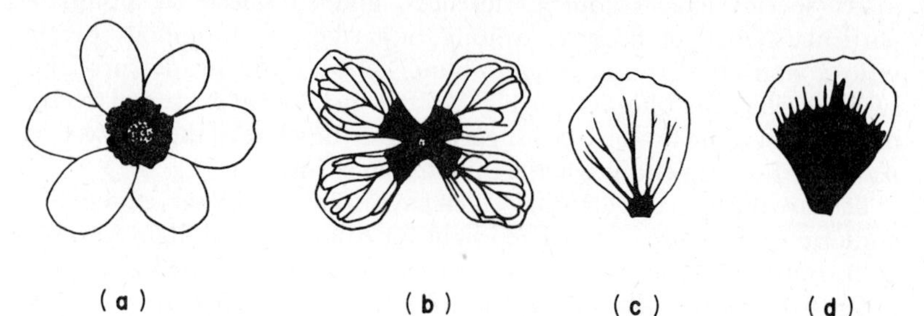

Fig. 5–11 The presence of nectar guides as seen under ultraviolet light. (**a**)=*Jasminum fruticans*; (**b**)=*Raphanus raphanistrum*; (**c**)=*Oenothera fruticosa*; (**d**)=*Oenothera albicans*. (Redrawn from KUGLER, H. (1963). *Planta*, 59, 307.)

6 Suggestions for Practical Work

6.1 Construction of a simple tube solarimeter

(Courtesy of D. Whale, Botany School, Oxford)

A solarimeter measures light intensity indirectly by sensing the temperature difference between black and white sectors which results from a differential absorption of the radiant flux incident on them. Solarimeters may be used to find correlation for horizontal and vertical profiles of light intensity with changes in species composition, abundance or distribution. Seasonal comparisons may be made between different types of woodland, across a woodland boundary, or within a tall grass community. Solarimeters may also be used in experiments on light requirements for seed germination, shade tolerance, etc.

Constantan wire is wound on a perspex former, or one of similar material, with the winding direction reversed midway along the length of the perspex (Fig. 6–1(c)). Although it is possible to wind the perspex by hand it is tedious and access to a metal-working lathe is desirable. Thermocouples are formed by copper plating all the constantan except the area indicated in Fig. 6–1(c) on the upper surface. This area is painted with Durofix or nail varnish prior to immersion in the plating bath (200 g $CuSO_4$: 50 g H_2SO_4 : 1000 g H_2O). A 12V battery may be used to provide a plating current of approximately 15–30 mA for 30 mins. Depending on the resistance of the winding it may be necessary to include a series resistance or to wire the formers in parallel to achieve this current range. To ensure an even deposit of plating secure some bare copper wire over the constantan and strain it to ensure close contact by bowing the perspex.

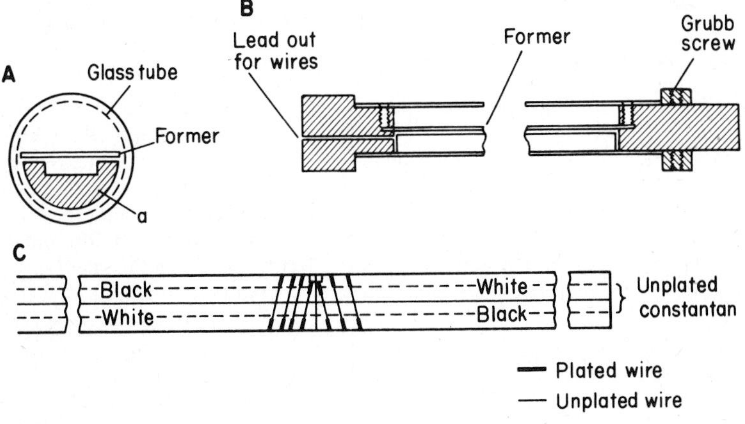

Fig. 6–1 Construction of a simple tube solarimeter. See text for details.

Paint the perspex with black and white matt paints as indicated in Fig. 6–1(c) leaving the underside white. Solder two lead wires to the constantan and mount the completed former in a glass tube taking care that the sides do not touch the glass. The formers are secured with perspex holders, as in Fig. 6–1(b) (or simply use rubber bungs with slits cut into them) and a seal effected using silicone rubber. Prior to sealing the tube introduce some silica gel into the trough which carries the lead wires. This latter piece ('a' in Fig. 6–1(a)) serves to support and shield the underside of the perspex former from radiation. The tube may be supported using terry clips.

Using 200 turns of 42 swg wire on a perspex former of $22.9 \times 1.3 \times 0.2$ cm readings up to 6 mV may be obtained in bright sun; steady readings may be attained within 30 seconds. The circuit shown in Fig. 6–2 will amplify the signal for use with a pH meter (scale 0–700 mV, impedance $10^{12}\Omega$) as recorder. Calibration may be effected relative to any available light meter such as a photographic exposure meter. If only a single solarimeter is available profiles should be measured under conditions of near uniform irradiance (preferably on an evenly overcast day) so that rapid flux changes do not result in spurious readings.

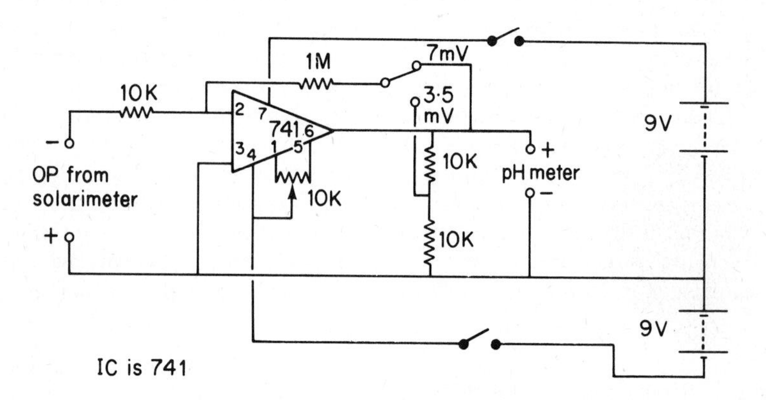

Fig. 6–2 Circuit diagram for use with a simple tube solarimeter. All resistances are 0.5 W, tolerance ±2% with low temperature coefficients (e.g. thick film). Connect the solarimeter, select the range, switch on and set zero using the 10K potentiometer whilst the solarimeter is shielded from any light. Remove the shield, allow the reading to stabilize (30 secs for the type described here) and then record it. The pH meter must be used in its millivoltmeter mode and its asymmetry control must be adjusted prior to effecting the dark adjustment above. (Courtesy of D. T. Smith, Clarendon Laboratory, Oxford.)

6.2 Germination

Sow 20 seeds of each of the experimental species on moist filter paper in two petri dishes. Place one dish in the dark; leave the other in the light, maintaining both at the same temperature (range 20–25°C). Examine each day, using a green safelight to see the dark set, and count the numbers of seeds which have germinated (emergence of radicle) at specified intervals. List seeds which are neutral, light-requiring or dark-requiring. If red and far-red filters are available test for the possible involvement of phytochrome (see Table 1 and pp. 27–31).

Suitable species to investigate could include *Lactuca sativa* (different varieties), *Sinapis alba, Viscum album, Epilobium* spp., *Oenothera* spp., *Lythrum salicaria, Allium* spp. *Phacelia*, spp., *Ranunculus sceleratus*, *Rigella* spp., and a selection of agriculturally important crop plants and weeds.

6.3 Etiolation

Soak seeds of French bean (*Phaseolus vulgaris*) in distilled water for 5 hours. Remove the testa from each seed and immediately plant the seeds in moist soil or sand in two flats. Place one flat in a light-tight box and the other flat in the light so that the temperature for both is about 24°C. Leave the seeds to develop for 7–10 days. The soil must be kept damp during this period though the dark-grown plants may need little water. However, the plants in the dark must not be exposed to the light even for a few seconds, so any watering or preliminary examination must be carried out using a green safelight. After 7–10 days compare the light- and dark-grown plants. Note the difference in colour between the two groups of plants. If a spectrophotometer is available measure the absorption spectrum of an 80% acetone extract of leaf tissue from each. Measure leaf size and internode length. Cut leaf sections by hand and examine them in the light microscope. Estimate the relative sizes of cells in both sets of plants, using a micrometer eyepiece. It may be interesting to stain some sections with I_2/KI solution to show starch.

6.4 Flowering

Into 8 cm pots plant seeds of 'Charm' or Korean chrysanthemums, but avoid the new early-flowering varieties. When large enough transfer to 13 cm pots, keep warm and moist but well ventilated in a greenhouse. When the plants are established arrange that one set of pots is kept in complete darkness for 16 continuous hours during each 24-hour period. Arrange other sets to be in complete darkness for shorter periods (8–15 hours) per 24 hours. Keep all treatments at the same temperature. Note which plants produce flower buds and how many are formed. (Refer to pp. 59–64 when interpreting your results.)

Further Reading

General References

LEOPOLD, A. C. (1964). *Plant Growth and Development*. McGraw-Hill.
SMITH, H. (1975). *Phytochrome and Photomorphogenesis*. McGraw-Hill.
SMITH, H. (ed.) (1976). *Light and Plant Development*. Butterworth.
VINCE–PRUE, D. (1975). *Photoperiodism in Plants*. McGraw-Hill.
WHITTINGHAM, C. P. (1971). *Photosynthesis*. Oxford Biology Readers, Oxford University Press.

Studies in Biology Series

HALL, D. O. and RAO, K. K. (1977). *Photosynthesis*. No. 37, Edward Arnold, London.
KENDRICK, R. E. and FRANKLAND, B. (1976). *Phytochrome and Plant Growth*. No. 68. Edward Arnold, London.
PHILLIPSON, J. (1966). *Ecological Energetics*. No. 1. Edward Arnold, London.
VILLIERS, T. A. (1975). *Dormancy and Survival of Plants*. No. 57. Edward Arnold, London.

Symposium Volumes

British Ecological Society Symposium 6 (1966). *Light as an Ecological Factor I*.
British Ecological Society Symposium 16 (1975). *Light as an Ecological Factor II*.